TEMPORAL MODALITIES IN ARABIC LOGIC

FOUNDATIONS OF LANGUAGE

SUPPLEMENTARY SERIES

VOLUME 2

THE HUMANITIES PRESS / NEW YORK

NICHOLAS RESCHER

TEMPORAL MODALITIES IN ARABIC LOGIC

D. REIDEL PUBLISHING COMPANY / DORDRECHT - HOLLAND

SOLE DISTRIBUTORS FOR U.S.A. AND CANADA
THE HUMANITIES PRESS / NEW YORK

Library of Congress Catalog Card Number 67–12055

1967

Printed in The Netherlands by D. Reidel, Dordrecht

Its author dedicates this monograph
in cordial friendship to his colleague
Alan Ross Anderson

PREFACE

The aim of this monograph is to expound the conceptions of temporalized modality at issue in various Arabic logical texts. I claim to have been able to make good logical sense of doctrines of which even the later Arab logicians themselves came to despair. In the process, a substantially new area of the history of logic has come into a clear view.

I am indebted to Anne Cross (Mrs. Michael) Pelon and especially Mr. Bas van Fraassen for assistance in the research. Miss Dorothy Henle merits my thanks for preparing the difficult typescript for the printer and helping to see the book through the press. Also, I am grateful to the Editors of *Foundations of Language* for inviting inclusion of the monograph in the *Supplementary Series* of the journal.

The present work is part of a series of studies of Arabic contributions to logic supported by research grants from the National Science Foundation. It affords me much pleasure to record my sincere thanks for this assistance.

Pittsburgh, January 1966 NICHOLAS RESCHER

CONTENTS

1 INTRODUCTION

In the wake of A. N. Prior's book on *Time and Modality*[1] an active interest has sprung up among logicians in the logical theory of chronological propositions generally, and particularly in the relationships that obtain between such propositions and modal concepts. This phenomenon is not surprising, because the issue is one that ramifies widely into various topics of logico-philosophical interest: the theory of tensed discourse, the problem of determinism, and the puzzle of future contingency, among others. The modern discussions have gone forward wholly oblivious to the fact that medieval Arabic logicians had given extensive attention to the development of a theory of temporal modalities, and had developed an extensive and subtle machinery for dealing with problems in this area. The aim of the present monograph is one of "intellectual archeology" – to present the Arabic contributions to this branch of logic in such a way that their linkage with ideas and concepts of present-day interest can be assessed and appreciated.

2 SOURCES

The principal basis for our discussion is "The Sun Epistle" *Al-Risālah al-shamsiyyah* of the thirteenth-century Persian philosopher-scientist al-Qazwīnī al-Kātibī[2] (ca. 1220–1276 or 1292). Not only is this work one of the few that treats of our problem in significant detail, but it is one of the very few Arabic logic treatises to have been put into a European language, as follows: Aloys Sprenger, *Dictionary of the Technical Terms Used in the Sciences of the Musulmans*, Part 2 (Calcutta, 1862). Appendix I (1862) on "The Logic of the Arabians" gives an Arabic text edition and a (slightly incomplete) English translation of our treatise. This translation has a serious shortcoming from the standpoint of our present concerns. As the translator explains, certain parts of the treatise (to wit the following sections Engl. 68–70 = Ar. 66–68; Engl. 72–74 = Ar. 70–72; Engl. 84–86 = Ar. 81–84) "are omitted in the translation because they contain details on modals which are of no interest. The last-named four paragraphs [dealing with modal syllogisms] are also omitted in most Arabic textbooks on Logic, and are not studied in Mohammedan

[1] Oxford, The Clarendon Press, 1957.

[2] For this Arabic logician – as well as all others to be mentioned here – see the biobibliographical register in N. Rescher, *The Development of Arabic Logic* (Pittsburgh, 1964); see pp. 203–204.

Schools." An English translation of these sections of the text is given in Appendix A below. No matter how difficult or boring this material proved for the Muslim schoolmaster, it is of the greatest relevance for our interests.

It goes perhaps without saying that, although we base our discussion in the main upon the treatise of al-Qazwīnī al-Kātibī, the exposition will draw upon a wider range of Arabic logicians, including such more prominent figures as Avicenna and Averroes. In particular, we shall see that al-Qazwīnī al-Kātibī follows and draws upon Avicenna. Details of this reliance will be obvious at many points to anyone who compares al-Qazwīnī's treatment with the corresponding treatment of Avicenna's *Kitāb al-ishārāt wa-'l-tanbīhāt*, which is fortunately accessible to European scholars in an excellent French version: A. M. Goichon (tr.), *Ibn Sīnā: Livre des Directives et Remarques* (Beyrouth and Paris, 1951). Al-Qazwīnī's dependence upon Avicenna is strikingly evidenced by the close parallelism in organization, mode of treatment, and substance between the two treatises.

3 BACKGROUND

The logic of modality was a relative latecomer to the domain of Arabic logic. Only after the time of Abū Bishr Mattā ibn Yūnus (ca. 870–940), translator of *Posterior Analytics*, did the study of Aristotle's modal syllogistic come to be taken up.[3] In its wake, interest in other Greek ideas regarding modality sprung up, and in the 10th century there was a lively polemic in the School of Baghdad – whose principal figure was al-Fārābī (ca. 873–950) – against Galen's views on modality, especially his rejection of the modality of possibility.[4] That mainstream of Arabic logic which stayed within the tradition of the School of Baghdad, culminating in Averroes (1126–1198), always remained closely faithful to Aristotelian views.[5] On the other hand, the influence of Galen, and especially – perhaps through his mediation – of the Stoics, made significant headway in that part of the Arabic logical tradition whose fountain-head was Avicenna.[6] There can be little doubt that the theory of

[3] 'Al-Fārābī on Logical Tradition' in N. Rescher, *Studies in the History of Arabic Logic* (Pittsburgh, 1963), pp. 21–27.

[4] See N. Rescher, *The Development of Arabic Logic* (op.cit.), p. 43.

[5] See, for example, 'Averroes' *Quaesitum* on Assertoric (Absolute) Propositions' in N. Rescher, *Studies in the History of Arabic Logic* (op.cit.), pp. 91–105. The extent to which the Spanish Muslim logicians were faithful to Aristotle is exemplified by the treatment of modal syllogistic by Abū 'l-Ṣalt (1068–1134). See 'Abū-'l-Ṣalt of Denia on Modal Syllogisms', *ibid.*, pp. 87–90.

[6] See, for example, 'Avicenna on the Logic of "Conditional" Propositions', *ibid.*, pp. 76–86. On the general phenomenon see N. Rescher, *The Development of Arabic Logic* (op.cit.), pp. 50 ff., and *idem, Galen and the Syllogism* (Pittsburgh, 1966), pp. 4–8.

temporal modalities with which we shall be dealing in the present monograph goes back, either entirely, or at any rate in its original and generative impetus, to Greek, and above all to Stoic logic. We shall, however, postpone any detailed consideration of the cognate conceptions of Megarian and Stoic until Section 13, after our examination of the Arabic materials has been completed.

The theory of temporalized modalities was developed by the Arabs within an Aristotelian setting, against the backdrop of a concept of demonstrative science insisting upon explanations in terms of premisses that are necessary in the sense of comprising constantly operative causes referring to what *always* happens.[7] Consider, for the sake of illustration, Avicenna's statement: "The mind is not at all repelled by the statement, "when Zayd moved his hand, the key moved", or "Zayd moved his hand, then the key moved." The mind is repelled, however, by the statement, "when the key moved, Zayd moved his hand", even though it is [rightly] said, "when the key moved, *we knew that* Zayd moved his hand." The mind, despite the temporal coexistence of the two movements, assigns a (causal) priority for one, a posteriority for the other. For it is not the existence of the second movement that causes the existence of the first; it is the first movement that causes the second."[8] Its rooting in the fertile soil of the concepts of causality and necessity provided for the Arabs the impetus to preoccupation with the logical theory of temporally modalized propositions.

4 FUNDAMENTAL IDEAS

The theory of modal propositions is superadded to or superimposed upon a basic theory of absolute, unmodalized, categorical propositions (of the standard A, E, I, and O type). Let us here use the variables

$$A, B, C, \ldots$$

to stand for such categorical propositions.

When a proposition of this sort is modalized, it bears a modal qualifier such as "necessarily", "perpetually", "non-necessarily", or "non-perpetually" (§ 48).[9]

[7] Cf. M. E. Marmura, 'Ghazali and Demonstrative Science', *Journal of the History of Philosophy* **3** (1965) 183–204.

[8] *Ibn Sīnā, Al-Shifā': al-Ilāhīyāt (Metaphysics)* (ed. C. G. Anawati, S. Dunya, and S. Zayd, revised by M. Madkur), 2 vols (Cairo, 1960), vol. I, p. 165.

[9] We adopt the practice of citing in this way the section-numbering adopted in the Sprenger *translation* of our treatise. His numbering of paragraphs in the English version unaccountably differs in places from that used for the Arabic text.

The idea of modality is introduced in § 48 in the following terms: "The relation of the predicates to the subjects must – be they affirmative or negative – have a certain qualification such as "necessarily", "perpetually", "non-necessarily", or "non-perpetually". Such a qualification is called the *materia* (*maddah*) of the proposition, and the word expressing it, is called the *mode* (*jihah*) of the proposition." In Avicenna's *Kitāb al-ishārāt wa-'l-tanbīhāt*[10] the same point is put as follows: "By the *materia* [of a proposition] we understand the three states upon which, in the affirmation rests the truth of these three words ["necessary", "possible", "impossible"] when they are used explicitly. (p. 134) A correspondence theory of modality appears to be at work here: there is an objective kind of fact *de re* correspondence which constitutes the truth-ground of the modalized proposition *de dicto*. Here and elsewhere (§§ 40, 41, 46) it is maintained that this modality-underwriting fact is one with respect to the sort of relation – and preeminently temporal relation – which obtains between the subject and the predicate.

In the explanation of various modalities, various technical ideas are required as special instruments.

One item of technical equipment is a bracket operator to indicate the subject of a categorical proposition so that

$$[A]$$

represents the subject of (the proposition) A.

Another needed item of technical equipment is an existence operator. This will be represented by 'E!', so that "E!X" says "X exists", and

$$E![A]$$

to be construed as "the essence of the subject of A exists" or perhaps as "the subject of A is actually exemplified". It should be noted that "$E![A]$" is equivalent with "$E![\sim A]$", where $\sim A$ is, of course, the contradictory of A.

5 BASIC MODAL RELATIONS

We shall need to work with four basic modal relationships:

1. $(A \mid \Box \mid B)$

for "A is necessarily true whenever B is true." (We shall call this relationship that of *chronologically constant necessity-correlation*.)

2. $(A \mid \forall t \mid B)$

[10] I cite this from the French version of A. M. Goichon, already mentioned above.

for "A is true whenever B is true." (We shall call this relationship that of *chronologically constant correlation.*)

$$3.\ (A \mid \exists t \mid B)$$

for "A is true at some time that B is true" (*chronologically occasional correlation*).

$$4.\ (A \mid \Diamond \mid B)$$

for "A is possible at some time that B is true" (*chronologically occasional possibility-correlation*).

To provide a fuller and more precise explication of the ideas at issue here let us introduce somewhat more symbolic machinery to wit t, t', t'', etc., as variables for *times*; $T_t(A)$ for "A is true at the time t"; $N_t(A)$ for "A is necessarily true at the time t"; P_t for "A is possibly true at the time t"; $\rightarrow$ for entailment or (strict) implication. Now we can construe the four aforementioned relationships as follows (with "iff" as abbreviation for "if and only if"):

$$1.\ (A \mid \Box \mid B) \text{ iff } (\forall t)\ [T_t(B) \rightarrow N_t(A)]$$
$$2.\ (A \mid \forall t \mid B) \text{ iff } (\forall t)\ [T_t(B) \rightarrow T_t(A)]$$
$$3.\ (A \mid \exists t \mid B) \text{ iff } (\exists t)\ [T_t(A)\ \&\ T_t(B)]$$
$$4.\ (A \mid \Diamond \mid B) \text{ iff } (\exists t)\ [P_t(A)\ \&\ T_t(B)]$$

If we accept the plausible principle that what is possible under given circumstances at one time is possible under those circumstances at all times, we could indifferently construe (4) as "A is possible at all times that B is true."

We could clearly adopt only two of these four relationships as primitive ideas, introducing the other two by means of the duality-equivalences

$$\sim(A \mid \Box \mid B) \text{ iff } (\sim A \mid \Diamond \mid B)$$
$$\sim(A \mid \forall t \mid B) \text{ iff } (\sim A \mid \exists t \mid B)$$

Moreover, it is clearly intended that we should have:

$$\text{If } (A \mid \Box \mid B), \text{ then } (A \mid \forall t \mid B)$$ [11]

This, of course, is an immediate consequence of the foregoing constructions. And then we also obtain derivatively

$$\text{If } (A \mid \exists t \mid B), \text{ then } (A \mid \Diamond \mid B)$$

The affirmativeness or negativeness of a complex of the form $(A \mid - \mid -)$ is to be considered as determined by that of A.

Thus, to give a specific example of the working of these ideas

$$(\text{All } S \text{ is } P \mid \forall t \mid E![\text{All } S \text{ is } P])$$

[11] Compare St. Thomas Aquinas: "Est igitur Deus aeternus: cum omne necessarium per se sit aeternum" (*Summa Contra Gentiles*, I, c. 15).

means: "All S is P as long as the essence of S exists, i.e., as long as there are S's."

This locution "as long as the essence of the subject exists" as it is used in §§ 50–51 appears to mean "throughout the entire history of the existence of the subject". For example, the modalized proposition "It is necessary that every man is an animal" is explained as meaning that the predicate of being an animal pertains of necessity to man as long as the essence of the subject exists – i.e., as long as there are men. In § 25 universals are grouped into three classes – physical, logical, and mental – with "animal" as the paradigmatic physical universal[12] said to have the locus of its existence in its instances, that is, in the existence of animals.

It is by means of – not the notations but – the *concepts* here introduced that our author defines the modal "qualifiers" characterizing "the relation of the predicates to the subjects" (§ 48) in modalized categorical propositions.

6 ENUMERATION OF MODAL PROPOSITIONS I. SIMPLE MODALITIES

There are six types of simple modal propositions (enumerated in § 50). Their character is as follows:

1. *Type* N_A: *The Absolute Necessary*

With respect to the *absolute necessary*, the text and examples of § 50.1 make it clear that this modalization of a categorical proposition P is to be construed as amounting to

$$(A \,/\, \Box \,/\, E![A]$$

that is to say, as asserting that A is necessarily the case as long as the essence of its subject exists (i.e., is exemplified). The text puts the matter as follows, that the proposition so modalized: "pronounces that the predicate is affirmed or denied [of an instance] of the subject of necessity as long as the essence of the subject exists [in this instance]." (§ 50.1) The examples given are: *Necessarily: Every man is an animal* and *Necessarily: No man is a stone.*

2. *Type* R_A: *The Absolute Perpetual*

The *absolute perpetual* proposition is to be construed as

$$(A \,/\, \forall t \,/\, E![A])$$

[12] The other two types of universals are ruled out of discussion as irrelevant.

that is to say, as asserting that A is perpetually the case as long as – i.e., is always true whenever – the essence of its subject exists (i.e., is exemplified). The proposition so modalized: "pronounces that the predicate is affirmed or denied of the subject in perpetuity as long as the essence of the subject exists." (§ 50.2) The examples given for N_A are said to be applicable here also, so that we have it that: *Perpetually: Every man is an animal* and *Perpetually: No man is a stone.*

3. *Type C_G: The General Conditional (Also: Absolute Temporal)*

The *general conditional* proposition is to be construed as

$$(A \mid \Box \mid C[A])$$

that is to say, as asserting that A is necessarily the case as long as its subject satisfies the condition C. It appears (at § 51.5) that this form of modality is also called the *absolute temporal.* The proposition so modalized: "pronounces that the predicate is affirmed or denied of necessity [of the subject] under the condition of [the obtaining of] a certain attribute of its subject." (§ 50.3) The examples given are: *Of (temporal) necessity: Every writer moves his fingers, as long as he writes,* and *Of (temporal) necessity: No writer keeps his fingers at rest, as long as he writes.*

4. *Type V_G: The General Conventional (Also: Absolute Spread)*

The *general conventional* proposition is to be construed as

$$(A \mid \forall t \mid C[A])$$

that is to say, as asserting that A is perpetually the case as long as its subject satisfies the condition C. The proposition so modalized: "pronounces that the predicate is affirmed or denied of the subject in perpetuity under the condition of [the obtaining of] a certain attribute of its subject." (§ 50.4) Again, the examples given in Section 3 for the general conditional proposition are said to serve here: *In (conditional) perpetuity: Every writer moves his fingers as long as he writes* and *In (conditional) perpetuity: No writer keeps his fingers at rest, as long as he writes.*

5. *Type A_G: The General Absolute*

The *general absolute* proposition is to be construed as

$$(A \mid \exists t \mid E![A]) \text{ or equivalently } \sim(\sim A \mid \forall t \mid E![\sim A])$$

that is to say that A is the case at some time, i.e., with its subject term construed to hold *at some time* for all its then concurrently existing instances. The proposition so modalized: "pronounces that the predicate is actually (or:

at some time) affirmed or denied of the subject." (§ 50.5) The examples cited are: *With general absoluteness: Every man is breathing* and *With general absoluteness: Every man is not breathing.*

Note that – as is claimed at the end of § 51 – the general absolute A_G is a contradictory of the absolute perpetual R_A.

6. *Type P_G: The General Possible*

The *general possible* proposition has the form

$$\sim(\sim A \;/\; \Box \;/\; E![\sim A]) \text{ or equivalently } (A \;/\; \Diamond \;/\; E![A])$$

that is to say that A is possible as long as (i.e., whenever) the subject of A exists. The proposition so modalized: "pronounces that there is no absolute necessity that what is contrary to the judgment should not be the case." (§ 50.6) The examples given are: *By a general possibility: All fires are cold* and *By a general possibility: Whatever is warm is not cold.*

Note again that – as is claimed at the end of § 51 – the general possible is a contradictory of the absolute necessary.

To summarize, then, we are presented with six basic types of temporally construed modalized propositions of the *simple* uncompounded variety, as follows:

Code[13]	Name	Construction	
1. N_A	Absolute necessary	$(A \;/\; \Box \;/\; E![A])$	"A is necessarily true whenever $E![A]$"
2. R_A	Absolute perpetual	$(A \;/\; \forall t \;/\; E![A])$	"A is perpetually true whenever $E![A]$"
3. C_G	General conditional [also "absolute temporal"]	$(A \;/\; \Box \;/\; C[A])$	"A is necessarily true whenever $C[A]$"
4. V_G	General conventional [also "absolute spread"]	$(A \;/\; \forall t \;/\; C[A])$	"A is perpetually true whenever $C[A]$"
5. A_G	General absolute	$(A \;/\; \exists t \;/\; E![A])$	"A is occasionally true when $E![A]$"
6. P_G	General possible	$(A \;/\; \Diamond \;/\; E![A])$	"A is possibly true when $E![A]$"

[13] Two types of modalities which would fit naturally into this framework are not mentioned at all, viz., (i) $(A \;/\; \exists t \;/\; C[A])$ and (ii) $(A \;/\; \Diamond \;/\; C[A])$. The reason for this apparent neglect is that these must be construed to amount to (5) and (6) respectively.

7 ENUMERATION OF MODAL PROPOSITIONS II. COMPOUND MODALITIES

Propositions are defined as being compound when "their verity is composed at the same time of an affirmation and a negation". (§ 49) Seven compound modal propositions are introduced in our text as follows:

7. *Type C_S: The Special Conditional*

The *special conditional* proposition is to be construed as

$$(A \mathbin{/} \Box \mathbin{/} C[A]) \ \& \ ({\sim}A \mathbin{/} \exists t \mathbin{/} E![{\sim}A])$$

It is, as our text puts it: "the same as the general conditional with the restriction that the relation of the subject [and predicate] is not perpetual in regard to the essence [of the subject]." (§ 51.1) In other words it is the combination of the general conditional with the corresponding negativized absolute perpetual, i.e., with the corresponding negative general absolute. (The affirmativeness or negativeness of the modal proposition turns on that of A.)

Two examples are given:

1. (Affirmative): Necessarily, every writer moves his fingers as long as he writes, but not perpetually (i.e., he does not move his fingers as long as he exists).

2. (Negative): Necessarily, no writer keeps his fingers at rest as long as he writes, but not perpetually (i.e., he does not keep his fingers at rest as long as he exists).

8. *Type V_S: The Special Conventional*

The *special conventional* proposition is to be construed as

$$(A \mathbin{/} \forall t \mathbin{/} C[A]) \ \& \ ({\sim}A \mathbin{/} \exists t \mathbin{/} E![{\sim}A])$$

As the text puts it, it is: "the same as the general conventional with the restriction that it [the relation of the subject and predicate] is not perpetual in regard to the essence [of the subject]." (§ 51.2) In other words, it is the combination of the general conventional with the denial of the corresponding absolute perpetual, i.e., with the corresponding negative general absolute. (The affirmativeness or negativeness of the modal proposition again turns on that of A.) We are told that the examples for the preceding case – the special conditional – serve for the present one also.

9. *Type $\bar{N}_E$: The Non-Necessary Existential*

The *non-necessary existential* proposition is to be construed as

$$(A \mathbin{/} \exists t \mathbin{/} E![A]) \ \& \ ({\sim}A \mathbin{/} \Diamond \mathbin{/} E![{\sim}A])$$

As the text puts it, it is: "the same as the general absolute with the restriction that it [the relation of the subject and predicate] is not one of necessity in regard to the essence [of the subject]." (§ 51.3) In other words, it is the combination of the general absolute with the denial of the corresponding absolute necessary, i.e., with the corresponding negative general possible.

Two examples are given:

1. (Affirmative): All men laugh (at some time – i.e., occasionally); but this is not so of necessity.

2. (Negative): No man laughs (at some time – i.e., occasionally); but this is not so of necessity.

10. *Type $\bar{R}_E$: The Non-Perpetual Existential*

The *non-perpetual existential* proposition is to be construed as

$$(A \mid \exists t \mid E![A]) \ \& \ ({\sim} A \mid \exists t \mid E![{\sim} A])$$

The text puts it that it is: "the same as the general absolute with the restriction of non-perpetuity in regard to the essence [of the subject]." (§ 51.4) It is thus the combination of the general absolute with the denial of the corresponding absolute perpetual, i.e., with the corresponding negative general absolute. As throughout its affirmativeness and negativeness hinges upon that of A. It is stated that the examples for the preceding case – that of the non-necessary existential – serve for the present one also.

11. *Type T: The Temporal*

The *temporal* proposition is to be construed as

$$(A \mid \Box \mid T[A]) \ \& \ ({\sim} A \mid \exists t \mid E![{\sim} A])$$

Here $T[A]$ represents a special condition on the subject $[A]$ – viz., that we have to do with a certain definite period of its existence. The text puts it that: "it pronounces that the predicate is affirmed or denied of the subject of necessity during a definite period of the existence of the subject, under the restriction of non-perpetuity in regard to the essence of the subject." (§ 51.5) In other words, it is the combination of the absolute temporal (so the text!) with the denial of the absolute perpetual, i.e., with the corresponding negative general absolute.

Two examples are given:

1. (Affirmative): Of necessity, an eclipse of the moon always takes place when the earth is placed between the sun and the moon; but this is not so perpetually.

2. (Negative): Of necessity, an eclipse of the moon never takes place when the earth is placed at right angles to the moon and the sun; but this is not so perpetually.

12. *Type S: The Spread*

The *spread* proposition is to be construed as

$$(A \mid \Box \mid S[A]) \;\&\; ({\sim}A \mid \exists t \mid E![{\sim}A])$$

Here $S[A]$ represents a special condition on the subject $[A]$ – viz., that have to do with some indefinite period of its existence.

Our text puts it that: "it pronounces that the predicate is affirmed or denied of the subject of necessity during an indefinite period of the existence of the subject, under the restriction of non-perpetuity in reference to the essence [of the subject]." (§ 51.6) In other words, it is the combination of the absolute spread [so the text!] corresponding negative general absolute.

Two examples are given:

1. (Affirmative): Of necessity, man breathes at certain times, but this is not so perpetually.
2. (Negative): Of necessity, man does not breathe at certain times, but this is not so perpetually.

13. *Type P_S: The Special Possible*

The *special possible* – also *contingent* – proposition is to be construed as

$$(A \mid \Diamond \mid E![A]) \;\&\; ({\sim}A \mid \Diamond \mid E![{\sim}A])$$

In other words, it is the combination of a general possible with the corresponding negativized general possible. As the text puts it, the special possible: "pronounces that there is no absolute necessity either for the existence or non-existence of the relation ... It is composed of two general possible propositions, one of which is affirmative and the other negative." (§ 51.7) As the text rightly observes, there is, in such propositions, no difference between the affirmative and the negative form. One example is given:

By special possibility: Every man is a writer.

To summarize, then, we are presented with seven basic types of temporally construed modalized propositions of the *compound* variety, as presented at the top of the next page.

Basically only five distinct modalities are involved here, (11) and (12) both capable of being viewed as special cases of (7).

A word must be said about the completeness of this enumeration. Let us dismiss (11) and (12) on the grounds adduced. Let it be that the negative component must take the form $({\sim}A \mid \exists t \mid E![{\sim}A])$ and $({\sim}A \mid \Diamond \mid E![{\sim}A])$. Then twelve theoretical possibilities confront us, as shown at the foot of page 12. There thus are, on first view, three seeming gaps in the enumeration, which

Code	Name	Construction: Affirmative Component	Construction: Negative Component
7. C_S	Special conditional	$(A\ /\ \Box\ /\ C[A])$	$(\sim A\ /\ \exists t\ /\ E![\sim A])$
8. V_S	Special conventional	$(A\ /\ \forall t\ /\ C[A])$	$(\sim A\ /\ \exists t\ /\ E![\sim A])$
9. $\bar{N}_E$	Non-necessary existential	$(A\ /\ \exists t\ /\ E![A])$	$(\sim A\ /\ \Diamond\ /\ E![\sim A])$
10. $\bar{R}_E$	Non-perpetual existential	$(A\ /\ \exists t\ /\ E![A])$	$(\sim A\ /\ \exists t\ /\ E![\sim A])$
11. T	Temporal	$(A\ /\ \Box\ /\ T[A])$	$(\sim A\ /\ \exists t\ /\ E![\sim A])$
12. S	Spread	$(A\ /\ \Box\ /\ S[A])$	$(\sim A\ /\ \exists t\ /\ E![\sim A])$
13. P_S	Special possible	$(A\ /\ \Diamond\ /\ E![A])$	$(\sim A\ /\ \Diamond\ /\ E![\sim A])$

we have labeled (α)–(γ) in the Remarks column. If, however, we adopt the principle that what is possible must happen sometimes, viz.,

$$\text{If } (A\ /\ \Diamond\ /\ -),\ \text{then } (A\ /\ \exists t\ /\ -)$$

we can at once eliminate (α) and (β) from the sphere of feasibility. Moreover, case (6a) reduces to (5b) when we put '$\sim A$' for 'A': thus (6a) does not represent an independent concatenation, and the omission of (γ) is apparent rather than real. This indicates the essential completeness of the system.

	Affirmative Component	Negative Component	Remarks
1a.	$(A\ /\ \Box\ /\ E![A])$	$(\sim A\ /\ \exists t\ /\ E![\sim A])$	cannot arise
1b.	ditto	$(\sim A\ /\ \Diamond\ /\ E![\sim A])$	cannot arise
2a.	$(A\ /\ \forall t\ /\ E![A])$	$(\sim A\ /\ \exists t\ /\ E![\sim A])$	cannot arise
2b.	ditto	$(\sim A\ /\ \Diamond\ /\ E![\sim A])$	(α)
3a.	$(A\ /\ \Box\ /\ C[A])$	$(\sim A\ /\ \exists t\ /\ E![\sim A])$	=case 7
3b.	ditto	$(\sim A\ /\ \Diamond\ /\ E![\sim A])$	cannot arise
4a.	$(A\ /\ \forall t\ /\ C[A])$	$(\sim A\ /\ \exists t\ /\ E![\sim A])$	=case 8
4b.	ditto	$(\sim A\ /\ \Diamond\ /\ E![\sim A])$	(β)
5a.	$(A\ /\ \exists t\ /\ E![A])$	$(\sim A\ /\ \exists t\ /\ E![\sim A])$	=case 10
5b.	ditto	$(\sim A\ /\ \Diamond\ /\ E![\sim A])$	=case 9
6a.	$(A\ /\ \Diamond\ /\ E![A])$	$(\sim A\ /\ \exists t\ /\ E![\sim A])$	(γ)
6b.	ditto	$(\sim A\ /\ \Diamond\ /\ E![\sim A])$	=case 13

We thus arrive at the following set of temporalized modalities whose coherence and completeness may now be regarded as established:

Name	Code	Construction	Type[14]
1. Absolute necessary	N_A	$(A\ /\ \Box\ /\ E![A])$	N(necessary)
2. Absolute perpetual	R_A	$(A\ /\ \forall t\ /\ E![A])$	A(assertoric)
3. General conditional	C_G	$(A\ /\ \Box\ /\ C[A])$	N
4. General conventional	V_G	$(A\ /\ \forall t\ /\ C[A])$	A
5. General absolute	A_G	$(A\ /\ \exists t\ /\ E![A])$	P(possible)
6. General possible	P_G	$(A\ /\ \Diamond\ /\ E![A])$	P
7. Special conditional	C_S	$(A\ /\ \Box\ /\ C[A])\ \&$ $(\sim A\ /\ \exists t\ /\ E![\sim A])$	N
8. Special conventional	V_S	$(A\ /\ \forall t\ /\ C[A])\ \&$ $(\sim A\ /\ \exists t\ /\ E![\sim A])$	A
9. Non-necessary existential	$\bar{N}_E$	$(A\ /\ \exists t\ /\ E![A])\ \&$ $(\sim A\ /\ \Diamond\ /\ E![\sim A])$	P
10. Non-perpetual existential	$\bar{R}_E$	$(A\ /\ \exists t\ /\ E![A])\ \&$ $(\sim A\ /\ \exists t\ /\ E![\sim A])$	P
11. Temporal	T	$(A\ /\ \Box\ /\ T[A])\ \&$ $(\sim A\ /\ \exists t\ /\ E![\sim A])$	N
12. Spread	S	$(A\ /\ \Box\ /\ S[A])\ \&$ $(\sim A\ /\ \exists t\ /\ E![\sim A])$	N
13. Special possible	P_S	$(A\ /\ \Diamond\ /\ E![A])\ \&$ $(\sim A\ /\ \Diamond\ /\ E![\sim A])$	P

8 RULES FOR CONTRADICTORIES

These rules, as stated in § 60 for the simple modal proposition are as shown at the top of page 14. It is curious that the last two cases (5) and (6) are wholly omitted in the discussion of § 60.

The rules for the contradictories of compound modal propositions are stated in §§ 61–62 as follows: The complex modal proposition consists of (1) a simple modal proposition as its base, and (2) a correspondingly negativized rider which is either a general absolute or a general possible. To form its contradictory we

[14] Cf. the discussion of Section 12 below.

Original		Contradictory	
Name	Construction	Construction	Name
1. Absolute necessary	$(A\ /\ \Box\ /\ E![A])$	$(\sim A\ /\ \Diamond\ /\ E![\sim A])$	General possible
2. Absolute perpetual	$(A\ /\ \forall t\ /\ E![A])$	$(\sim A\ /\ \exists t\ /\ E![\sim A])$	General absolute
3. General conditional	$(A\ /\ \Box\ /\ C[A])$	$(\sim A\ /\ \Diamond\ /\ E![\sim A])$	General possible*
4. General conventional	$(A\ /\ \forall t\ /\ C[A])$	$(\sim A\ /\ \exists t\ /\ E![\sim A])$	General absolute**
5. General absolute	$(A\ /\ \exists t\ /\ E![A])$	$(\sim A\ /\ \forall t\ /\ E![\sim A])$	Absolute perpetual
6. General possible	$(A\ /\ \Diamond\ /\ E![A])$	$(\sim A\ /\ \Box\ /\ E![\sim A])$	Absolute necessary

* Possible temporal in the text.
** Absolute temporal in the text: perhaps for $(\sim A/\Diamond/C[\sim A])$.
NOTE: In contradicting those propositions (conventional and conditional) which involve a special condition C, the procedure seems to be to replace this special condition by the more general existence-condition $E!$

i. Form the contradictory of its modal base, and
ii. Disjoin it with the contradictory of its rider.

Thus by this rule the contradictory of the non-perpetual existential

$$(A\ /\ \exists t\ /\ E![A])\ \&\ (\sim A\ /\ \exists t\ /\ E![\sim A])$$

would be

$$(\sim A\ /\ \forall t\ /\ E![\sim A])\ \vee\ (A\ /\ \forall t\ /\ E![A])$$

that is, a disjunction of two absolute perpetuals.[15]

9 CONVERSION (I.E., SIMPLE CONVERSION)

The treatment of conversion (*conversio simplex*) for categorical propositions – as described in § 64 – is, of course, routine and straightforward: "'Even conversion' is an expression which means that the first part [i.e., *term*] of a proposition is put second and the second part first, and that the truth and *quale* remain unaltered." (§ 64) The situation is as follows:

[15] And so our text in § 61.

	Original	Converse
A	All *S* is *P*	Some *P* is *S* ["limitation"]
E	No *S* is *P*	No *P* is *S*
I	Some *S* is *P*	Some *P* is *S*
O	Some *S* is not *P*	Some *P* is not *S* [INVALID]

In asking for the converse of a modalized proposition (A / – / –), what we are after is another modalized proposition (B / – / –) such that: (1) B is the converse of A, and (2) this second modalized proposition, involving B, is a valid consequence of the first, involving A.

Consider, for the sake of an example, the **E** proposition *No S is P* modalized as an absolute perpetual

$$(\text{No } S \text{ is } P \mid \forall t \mid E!S)$$

From this it immediately follows that

$$(\text{No } P \text{ is } S \mid \forall t \mid E!S)$$

Now if we are entitled to assume that S and P are "temporally coordinate" – i.e., that E!S whenever E!P, and conversely – we can infer

$$(\text{No } P \text{ is } S \mid \forall t \mid E!P)$$

which is the absolute perpetual for the converse of our original **E** proposition. This illustrates the process of conversion (i.e., simple conversion) among modalized propositions – and also illustrates its dependence upon an assumption of "temporal coordination".

The claims with respect to conversion made by our author – as presented in §§ 64–68 of the text – are summarized in Tables I–IV. It must be remarked that these tables embody some errors of omission and commission. For example, Table IV which claims, following § 68, that **O** propositions are convertible in certain modalities, is clearly incorrect. In view of the inherent complexity of the logical processes at issue, this is not surprising, especially in the absence of any sort of tabular or symbolic machinery. Our author's discussion of the conversion of modal propositions is in substance very close to – though not wholly identical with – that of Bar Hebraeus (also derived from Avicenna) as presented in a form conveniently accessible to the non-orientalist in H. F. Janssens, *L'Entretien de la Sagesse de Bar Hebraeus* (Liège and Paris, 1937; Bibliothèque de la Faculté de Philosophie et des Lettres de l'Université de Liège, fasc. LXXV), pp. 186–193. For details, the tabulation of the date in Table V on p. 20 should be consulted.

TABLE I

SIMPLE CONVERSION

A Propositions (A is All S is P)
Limited Conversion to I (Some P is S)

Original		Converse	
Name	Construction	Construction	Name
1. Absolute necessary	$(A / \Box / E![A]$	(Some P is S / $\Box$ / $C(P)$)	Absolute temporal* (§ 69)
2. Absolute perpetual	$(A / \forall t / E![A])$	ditto	Absolute temporal (§ 69)
3. General conditional	$(A / \Box / C[A])$	ditto	Absolute temporal (§ 69)
4. General conventional	$(A / \forall t / C[A])$	ditto	Absolute temporal (§ 69)
5. General absolute	$(A / \exists t / E![A])$	(Some P is S / $\exists t$ / $E!P$)	General absolute (§ 70)
6. General possible	$(A / \Diamond / E![A])$	? ("unknown")	(§ 70)
7. Special conditional	$(A / \Box / C[A])$ & $(\sim A / \exists t / E![A])$	(Some P is S / $\Box$ / $C(P)$) & (No P is S / $\exists t$ / $E!P$)	Special conditional (§ 69)
8. Special conventional	$(A / \forall t / C[A])$ & $(\sim A / \exists t / E![A])$	ditto	Special conditional (§ 69)
9. Non-necessary existential	$(A / \exists t / E![A])$ & $(\sim A / \Diamond / E![A])$	(No P is S / $\exists t$ / $E!P$)	General absolute (§ 70)
10. Non-perpetual existential	$(A / \exists t / E![A])$ & $(\sim A / \exists t / E![A])$	ditto	General absolute (§ 70)
11. Temporal	$(A / \Box / T[A])$ & $(\sim A / \exists t / E![A])$	ditto	General absolute (§ 70)
12. Spread	$(A / \Box / S[A])$ & $(\sim A / \exists t / E![A])$	ditto	General absolute (§ 70)
13. Special possible	$(A / \Diamond / E![A])$ & $(\sim A / \Diamond / E![A])$	? ("unknown")	(§ 70)

* Apparently the "absolute temporal" here is the general conditional, i.e., the temporal (no. 11) without the restriction.

TABLE II

SIMPLE CONVERSION

E Propositions (*A* is No *S* is *P*)

Original		Converse	
Name	Construction	Construction	Name
1. Absolute necessary	$(A\ /\ \Box\ /\ E![A])$	(No *P* is $S\ /\ \forall t\ /\ E!P$)	Absolute perpetual (§ 66)
2. Absolute perpetual	$(A\ /\ \forall t\ /\ E![A])$	ditto	Absolute perpetual (§ 66)
3. General conditional	$(A\ /\ \Box\ /\ C[A])$	(No *P* is $S\ /\ \forall t\ /\ C(P)$)	General conventional (§ 67)
4. General conventional	$(A\ /\ \forall t\ /\ C[A])$	ditto	General conventional (§ 67)
5. General absolute	$(A\ /\ \exists t\ /\ E![A])$	–	NONE (§ 65)
6. General possible	$(A\ /\ \Diamond\ /\ E![A])$	–	NONE (§ 65)
7. Special conditional	$(A\ /\ \Box\ /\ C[A])$ & $(\sim A\ /\ \exists t\ /\ E![A])$	(No *P* is $S\ /\ \forall t\ /\ C(P)$) & (Some *P* is $S\ /\ \exists t\ /\ E!P$)	Special conventional (§ 67)
8. Special conventional	$(A\ /\ \forall t\ /\ C[A])$ & $(\sim A\ /\ \exists t\ /\ E![A])$	ditto	Special conventional (§ 67)
9. Non-necessary existential	$(A\ /\ \exists t\ /\ E![A])$ & $(\sim A\ /\ \Diamond\ /\ E![A])$	–	NONE (§ 65)
10. Non-perpetual existential	$(A\ /\ \exists t\ /\ E![A])$ & $(\sim A\ /\ \exists t\ /\ E![A])$	–	NONE (§ 65)
11. Temporal	$(A\ /\ \Box\ /\ T[A])$ & $(\sim A\ /\ \exists t\ /\ E![A])$	–	NONE (§ 65)
12. Spread	$(A\ /\ \Box\ /\ S[A])$ & $(\sim A\ /\ \exists t\ /\ E![A])$	–	NONE (§ 65)*
13. Special possible	$(A\ /\ \Diamond\ /\ E![A])$ & $(\sim A\ /\ \Diamond\ /\ E![A])$	–	NONE (§ 65)

* *Q:* Is this one of the temporals of the first sentence of § 65? *A:* Yes – by elimination.

TABLE III

SIMPLE CONVERSION

I Propositions (*A* is Some *S* is *P*)

Original		Converse	
Name	Construction	Construction	Name
1. Absolute necessary	$(A\ /\ \Box\ /\ E![A])$	(Some *P* is *S* / $\Box$ / $C(P)$)	Absolute temporal* (§ 69)
2. Absolute perpetual	$(A\ /\ \forall t\ /\ E![A])$	ditto	Absolute temporal (§ 69)
3. General conditional	$(A\ /\ \Box\ /\ C[A])$	ditto	Absolute temporal (§ 69)
4. General conventional	$(A\ /\ \forall t\ /\ C[A])$	ditto	Absolute temporal (§ 69)
5. General absolute	$(A\ /\ \exists t\ /\ E![A])$	(Some *P* is *S* / $\exists t$ / $E!P$)	General absolute (§ 70)
6. General possible	$(A\ /\ \Diamond\ /\ E![A])$	? ("unknown")	(§ 70)
7. Special conditional	$(A\ /\ \Box\ /\ C[A])$ & $(\sim A\ /\ \exists t\ /\ E![A])$	(Some *P* is *S* / $\Box$ / $C(P)$) & (No *P* is *S* / $\exists t$ / $E!P$)	Special conditional (§ 69)
8. Special conventional	$(A\ /\ \forall t\ /\ C[A])$ & $(\sim A\ /\ \exists t\ /\ E![A])$	ditto	Special conditional (§ 69)
9. Non-necessary existential	$(A\ /\ \exists t\ /\ E![A])$ & $(\sim A\ /\ \Diamond\ /\ E![A])$	(Some *P* is *S* / $\exists t$ / $E!P$)	General absolute (§ 70)
10. Non-perpetual existential	$(A\ /\ \exists t\ /\ E![A])$ & $(\sim A\ /\ \exists t\ /\ E![A])$	ditto	General absolute (§ 70)
11. Temporal	$(A\ /\ \Box\ /\ T[A])$ & $(\sim A\ /\ \exists t\ /\ E![A])$	ditto	General absolute (§ 70)
12. Spread	$(A\ /\ \Box\ /\ S[A])$ & $(\sim A\ /\ \exists t\ /\ E![A])$	ditto	General absolute (§ 70)
13. Special possible	$(A\ /\ \Diamond\ /\ E![A])$ & $(\sim A\ /\ \Diamond\ /\ E![A])$	? ("unknown")	(§ 70)

* See the annotation to Table I above.

TABLE IV

SIMPLE CONVERSION

O Propositions (A is Some S is not P)

Original		Converse	
Name	Construction	Construction	Name
1. Absolute necessary	$(A\ /\ \Box\ /\ E![A])$	–	NONE
2. Absolute perpetual	$(A\ /\ \forall t\ /\ E![A])$	–	NONE
3. General conditional	$(A\ /\ \Box\ /\ C[A])$	–	NONE
4. General conventional	$(A\ /\ \forall t\ /\ C[A])$	–	NONE
5. General absolute	$(A\ /\ \exists t\ /\ E![A])$	–	NONE
6. General possible	$(A\ /\ \Diamond\ /\ E![A])$	–	NONE
7. Special conditional	$(A\ /\ \Box\ /\ C[A])$ & $(\sim A\ /\ \exists t\ /\ E![A])$	(Some P is not S / $\forall t$ / $C(P)$) & (All P is S / $\exists t$ / $E!P$)	Special conventional (§ 68)
8. Special conventional	$(A\ /\ \forall t\ /\ C[A])$ & $(\sim A\ /\ \exists t\ /\ E![A])$	ditto	Special conventional (§ 68)
9. Non-necessary existential	$(A\ /\ \exists t\ /\ E![A])$ & $(\sim A\ /\ \Diamond\ /\ E![A])$	–	NONE
10. Non-perpetual existential	$(A\ /\ \exists t\ /\ E![A])$ & $(\sim A\ /\ \exists t\ /\ E![A])$	–	NONE
11. Temporal	$(A\ /\ \Box\ /\ T[A])$ & $(\sim A\ /\ \exists t\ /\ E![A])$	–	NONE
12. Spread	$(A\ /\ \Box\ /\ S[A])$ & $(\sim A\ /\ \exists t\ /\ E![A])$	–	NONE
13. Special possible	$(A\ /\ \Diamond\ /\ E![A])$ & $(\sim A\ /\ \Diamond\ /\ E![A])$	–	NONE

TABLE V

THE SIMPLE CONVERSION OF MODAL PROPOSITIONS ACCORDING TO BAR HEBRAEUS[1]

Original		Converse	
A *Propositions and* **I** *Propositions*			
1. Absolute[2] (i.e., general absolute)	{ All *S* is *P* / Some *S* is *P* }	Some *P* is *S*	Absolute
2. Necessary (i.e., absolute necessary)	Some *S* is *P*	Some *P* is *S*	Absolute[3]
3. Special possible	{ All *S* is *P* / Some *S* is *P* }	Some *P* is *S*	General possible[4]
4. Perpetual (permanent)	(case not discussed)		
5. Temporal	(case not discussed)		
O *Propositions*			
None convertible at all.[5]			
E *Propositions*			
1. Absolute (i.e., general absolute)	No *S* is *P*	not convertible	
3. Necessary (i.e., absolute necessary)	No *S* is *P*	No *P* is *S*	Necessary[6]
3. Special possible	No *S* is *P*	not convertible	
4. Perpetual (permanent)	No *S* is *P*	No *P* is *S*	Perpetual
5. Temporal	No *S* is *P*	not convertible	

[1] From H. F. Janssens (ed.), *L'Entretien de la Sagesse de Bar Hebraeus* (Liège and Paris, 1937; Bibliothèque de la Faculté de Philosophie et des Lettres de l'Université de Liège, fasc. LXXV), pp. 186–193.

[2] This is apparently what we know from al-Qazwīnī al-Kātibī as the *general absolute* ($A \mid \exists t \mid E!\,[A]$), and not the "absolute" in the sense of the unmodalized or strictly assertoric.

[3] In al-Qazwīnī al-Kātibī, this is not the (general) absolute, but the "absolute temporal" (§ 69).

[4] According to al-Qazwīnī al-Kātibī, this conversion is "unknown" (§ 70).

[5] So far as the types of modality at issue are concerned, this agrees with al-Qazwīnī al-Kātibī. But compare Table IV above.

[6] According to al-Qazwīnī al-Kātibī (§ 66), this is not the (absolute) necessary, but the absolute *perpetual.*

10 C-CONVERSION (CONVERSION BY CONTRADICTION)

The logical process of *conversion by contradiction* is described in the following terms: "This expression ['conversion by contradiction'] means: to place the contradictory of the second part of a proposition first, and the first part, unaltered, second. The *quale* of the new proposition will be the opposite of the original proposition, but it will be equally true." (§ 71) The workings of this process may be indicated by the following tabulation:

	Original	C-Converse
A	All *S* is *P*	No non-*P* is *S*
E	No *S* is *P*	Some non-*P* is *S* ["limitation"]
I	Some *S* is *P*	Some non-*P* is not *S* [INVALID]
O	Some *S* is not *P*	Some non-*P* is *S*

The process of *C*-conversion appears to call for three steps: (1) change *S* to non-*P*, (2) change *P* to *S*, and (3) change the *quality* of the proposition: in short, the *C*-converse of a categorical proposition is its converted obverse. Thus in asking for the *C*-converse of a modalized proposition $(A / - / -)$ what we are after is another modalized proposition $(B / - / -)$ such that: (1) *B* is the converse of *A*, and (2) this second modalized proposition, involving *B*, is a valid consequence of the first, involving *A*.

Consider, for the sake of an example, the **A** proposition *All S is P* modalized as an absolute perpetual

$$(\text{All } S \text{ is } P \mid \forall t \mid E!S)$$

From this it immediately follows that

$$(\text{No non-}P \text{ is } S \mid \forall t \mid E!S)$$

Now if we are entitled to assume that *S* and *non-P* are "temporally coordinate" – i.e., that *E*!*S* whenever *E*!non-*P*, and conversely – we can infer

$$(\text{No non-}P \text{ is } S \mid \forall t \mid E!\text{non-}P)$$

which is the absolute-perpetual for the *C*-converse of our original **A** proposition. This illustrates the process of *C*-conversion (i.e., conversion by contradiction) among modalized propositions. (Again, it also illustrates the dependency of this process upon an assumption of "temporal coordination.")

The claims with respect to *C*-conversion made by our author – as presented in §§ 72–74 of the text – are summarized in Tables VI–IX. Again, the tabulation embodies some errors of omission and commission. For example, in some modalities **E** propositions should be *C*-convertible by limitation, whereas the *C*-convertibility of **I** propositions should always fail.

TABLE VI

C-CONVERSION FOR A-PROPOSITIONS

(Here A = All S is P)

Original		C-Converse	
Name	Construction	Construction	Name
1. Absolute necessary	$(A\ /\ \Box\ /\ E![A])$	(No non-P is S / $\forall t$ / $E!$non-P)	Absolute perpetual
2. Absolute perpetual	$(A\ /\ \forall t\ /\ E![A])$	ditto	Absolute perpetual
3. General conditional	$(A\ /\ \Box\ /\ C[A])$	(No non-P is S / $\forall t$ / C(non-P))	General conventional
4. General conventional	$(A\ /\ \forall t\ /\ C[A])$	ditto	General conventional
5. General absolute	$(A\ /\ \exists t\ /\ E![A])$	–	NONE
6. General possible	$(A\ /\ \Diamond\ /\ E![A])$	–	NONE
7. Special conditional	$(A\ /\ \Box\ /\ C[A])$ & $({\sim}A\ /\ \exists t\ /\ E![A])$	(Some non-P is not S / $\forall t$ / C(non-P)) & (Some non-P is not S / $\exists t$ / $E!$non-P)	Special conventional
8. Special conventional	$(A\ /\ \forall t\ /\ C[A])$ & $({\sim}A\ /\ \exists t\ /\ E![A])$	ditto	Special conventional
9. Non-necessary existential	$(A\ /\ \exists t\ /\ E![\mathrm{A}])$ & $({\sim}A\ /\ \Diamond\ /\ E![A])$	–	NONE
10. Non-perpetual existential	$(A\ /\ \exists t\ /\ E![A])$ & $({\sim}A\ /\ \exists t\ /\ E![A])$	–	NONE
11. Temporal	$(A\ /\ \Box\ /\ T[A])$ & $({\sim}A\ /\ \exists t\ /\ E![A])$	–	NONE
12. Spread	$(A\ /\ \Box\ /\ S[A])$ & $({\sim}A\ /\ \exists t\ /\ E![A])$	–	NONE
13. Special possible	$(A\ /\ \Diamond\ /\ E![A])$ & $({\sim}A\ /\ \Diamond\ /\ E![A])$	–	NONE

All data from § 72.

TABLE VII

C-CONVERSION FOR E-PROPOSITIONS

(Here A = No S is P)

Original		C-Converse	
Name	Construction	Construction	Name
1. Absolute necessary	$(A \mid \Box \mid E![A])$		No Convertibility*
2. Absolute perpetual	$(A \mid \forall t \mid E![A])$		
3. General conditional	$(A \mid \Box \mid C[A])$		
4. General conventional	$(A \mid \forall t \mid C[A])$		
5. General absolute	$(A \mid \exists t \mid E![A])$		
6. General possible	$(A \mid \Diamond \mid E![A])$		
7. Special conditional	$(A \mid \Box \mid C[A])$ & $(\sim A \mid \exists t \mid E![A])$		
8. Special conventional	$(A \mid \forall t \mid C[A])$ & $(\sim A \mid \exists t \mid E![A])$		
9. Non-necessary existential	$(A \mid \exists t \mid E![A])$ & $(\sim A \mid \Diamond \mid E![A])$		
10. Non-perpetual existential	$(A \mid \exists t \mid E![A])$ & $(\sim A \mid \exists t \mid E![A])$		
11. Temporal	$(A \mid \Box \mid T[A])$ & $(\sim A \mid \exists t \mid E![A])$		
12. Spread	$(A \mid \Box \mid S[A])$ & $(\sim A \mid \exists t \mid E![A])$		
13. Special possible	$(A \mid \Diamond \mid E![A])$ & $(\sim A \mid \Diamond \mid E![A])$		

* Some should go by limitation!

All data from § 74.

TABLE VIII

C-CONVERSION FOR I-PROPOSITIONS

(Here A = Some S is P)

Original		C-Converse	
Name	Construction	Construction	Name
1. Absolute necessary	$(A\ /\ \Box\ /\ E![A])$	–	NONE
2. Absolute perpetual	$(A\ /\ \forall t\ /\ E![A])$	–	NONE
3. General conditional	$(A\ /\ \Box\ /\ C[A])$	–	NONE
4. General conventional	$(A\ /\ \forall t\ /\ C[A])$	–	NONE
5. General absolute	$(A\ /\ \exists t\ /\ E![A])$	–	NONE
6. General possible	$(A\ /\ \Diamond\ /\ E![A])$	–	NONE
7. Special conditional	$(A\ /\ \Box\ /\ C[A])$ & $(\sim A\ /\ \exists t\ /\ E![A])$	(Some non-P is not $S\ /\ \forall t\ /\ C$(non-P)) & (All non-P is $S\ /\ \exists t\ /\ E!$non-P)	Special conventional*
8. Special conventional	$(A\ /\ \forall t\ /\ C[A])$ & $(\sim A\ /\ \exists t\ /\ E![A])$	ditto	Special conventional*
9. Non-necessary existential	$(A\ /\ \exists t\ /\ E![A])$ & $(\sim A\ /\ \Diamond\ /\ E![A])$	–	NONE
10. Non-perpetual existential	$(A\ /\ \exists t\ /\ E![A])$ & $(\sim A\ /\ \exists t\ /\ E![A])$	–	NONE
11. Temporal	$(A\ /\ \Box\ /\ T[A])$ & $(\sim A\ /\ \exists t\ /\ E![A])$	–	NONE
12. Spread	$(A\ /\ \Box\ /\ S[A])$ & $(\sim A\ /\ \exists t\ /\ E![A])$	–	NONE
13. Special possible	$(A\ /\ \Diamond\ /\ E![A])$ & $(\sim A\ /\ \Diamond\ /\ E![A])$	–	NONE

* Should actually fail.

All data from § 73.

TABLE IX

C-CONVERSION FOR O-PROPOSITIONS

(Here A = Some S is not P)

Original		C-Converse	
Name	Construction	Construction	Name
1. Absolute necessary	$(A\ /\ \Box\ /\ E![A])$	–	NONE
2. Absolute perpetual	$(A\ /\ \forall t\ /\ E![A])$	–	NONE
3. General conditional	$(A\ /\ \Box\ /\ C[A])$	–	NONE
4. General conventional	$(A\ /\ \forall t\ /\ C[A])$	–	NONE
5. General absolute	$(A\ /\ \exists t\ /\ E![A])$	–	NONE
6. General possible	$(A\ /\ \Diamond\ /\ E![A])$	–	NONE
7. Special conditional	$(A\ /\ \Box\ /\ C[A])$ & $(\sim A\ /\ \exists t\ /\ E![A])$	(Some non-P is S / $\Box$ / $E!S$) & (All S is P / $\exists t$ / $E!S$)	Temporal
8. Special conventional	$(A\ /\ \forall t\ /\ C[A])$ & $(\sim A\ /\ \exists t\ /\ E![A])$	ditto	Temporal
9. Non-necessary existential	$(A\ /\ \exists t\ /\ E![A])$ & $(\sim A\ /\ \Diamond\ /\ E![A])$	(Some non-P is S / ∃t / E!non-P)	General absolute
10. Non-perpetual existential	$(A\ /\ \exists t\ /\ E![A])$ & $(\sim A\ /\ \exists t\ /\ E![A])$	ditto	General absolute
11. Temporal	$(A\ /\ \Box\ /\ T[A])$ & $(\sim A\ /\ \exists t\ /\ E![A])$	ditto	General absolute
12. Spread	$(A\ /\ \Box\ /\ S[A])$ & $(\sim A\ /\ \exists t\ /\ E![A])$	ditto	General absolute
13. Special possible	$(A\ /\ \Diamond\ /\ E![A])$ & $(\sim A\ /\ \Diamond\ /\ E![A])$	–	NONE

All data from § 74.

11 MODAL SYLLOGISMS

Rules for determining the modality of the conclusion of syllogisms (that is, assertorically valid syllogisms) with modalized premisses are presented in §§ 84–87. Take for example the AAA-1 (*Barbara*) syllogism with a necessary absolute major and a general conditional minor:

N_A: All M is P	(All M is P / $\Box$ / $E!M$)
C_G: All S is M	(All S is M / $\Box$ / $C[S]$)
? : All S is P	(All S is P / ? / ?)

Note that when $C[S]$ we must have it that $E!S$, and that therefore – since *All S is M* – the condition $E!M$ must be satisfied. Thus the given premisses warrant the conclusion

(All S is P / $\Box$ / $C[S]$)

which is a general conditional. Again, it is clear by analogous reasoning that an AAA-1 (Barbara) syllogism with two absolute perpetual premisses must yield an absolute perpetual conclusion. These examples show the process of syllogistic inference among modalized propositions.

The claims with respect to modal syllogisms made by our author – as presented in §§ 84–87 of the text – are summarized in Tables X–XIII. Again, it must be recognized that the tabulation embodies various errors of omission and commission. Considering the intricacy and complexity of the reasoning even with the aid of modern symbolic techniques, this fact should not be viewed as surprising. Indeed, one cannot but admire the fortitude of the old logicians in tackling these problems with the crude methods at their disposal.

12 AVICENNA AS THE SOURCE OF AL-QAZWĪNĪ AL-KĀTIBĪ'S LOGIC OF MODALITY

In Avicenna's *Kitāb al-ishārāt*[16], a treatment of modalities is given which may instructively be compared with that of our treatise. As regards the various kinds of necessity, the situation is as follows:

Avicenna	*al-Qazwīnī*
(P. 134) Every proposition is *either* absolute [i.e., unmodalized], of general application,	§ 48 [the absolute or unmodalized]

[16] *Kitāb al-ishārāt wa-'l-tanbīhāt*, tr. A. M. Goichon, *Ibn Sīnā: Livre des Directives et Remarques* (Beyrouth and Paris, 1951).

TABLE X

THE MODALITY OF VALID SYLLOGISMS IN THE FIRST FIGURE

Minor \ Major	1 N_A	2 R_A	3 C_G	4 V_G	5 A_G	6 P_G	7 C_S	8 V_S	9 $\bar{N}_E$	10 $\bar{R}_E$	11 T	12 S	13 P_S
1. N_A Absolute necessary				As along the top of the table									
2. R_A Absolute perpetual				As along the top of the table									
3. C_G General conditional	C_G	C_G	C_G	C_G	C_G	C_G	C_S	C_S	C_G	C_G	C_G	C_G	C_G
4. V_G General conventional	V_G	V_G	V_G	V_G	V_G	V_G	V_S	V_S	V_G	V_G	V_G	V_G	V_G
5. A_G General absolute				As along the top of the table									
6. P_G General possible				As along the top of the table									
7. C_S Special conditional	C_S	C_S	C_G	C_G	C_S	C_S	C_S	C_S	C_S	C_S	C_S	C_S	C_S
8. V_S Special conventional	V_S	V_S	V_G	V_G	V_S	V_S	V_S	V_S	V_S	V_S	V_S	V_S	V_S
9. $\bar{N}_E$ Non-necessary existential				As along the top of the table									
10. $\bar{R}_E$ Non-perpetual existential				As along the top of the table									
11. T Temporal				As along the top of the table									
12. S Spread				As along the top of the table									
13. P_S Special possible				As along the top of the table									

NOTE: The table indicates the modality of the conclusion for the (assertorically) valid syllogisms.

All data from § 83.

TABLE XI

THE MODALITY OF VALID SYLLOGISMS IN THE SECOND FIGURE*

Minor \ Major	1 N_A	2 R_A	3 C_G	4 V_G	5 A_G	6 P_G	7 C_S	8 V_S	9 $\bar{N}_E$	10 $\bar{R}_E$	11 T	12 S	13 P_S
1. N_A Absolute necessary	R_A	R_A	R_A	R_A	R_A	R_A	R_A	R_A	R_A	R_A	R_A	R_A	R_A
2. R_A Absolute perpetual	R_A	R_A	R_A	R_A	R_A		R_A	R_A	R_A	R_A	R_A	R_A	
3. C_G General conditional	**						C_G	C_G					
4. V_G General conventional							V_G	V_G					
5. A_G General absolute							A_G	A_G					
6. P_G General possible	R_A		P_G				P_G						
7. C_S Special conditional							C_G	C_G					
8. V_S Special conventional							V_G	V_G					
9. $\bar{N}_E$ Non-necessary existential							A_G	A_G					
10. $\bar{R}_E$ Non-perpetual existential							A_G	A_G					
11. T Temporal							C_G	C_G					
12. S Spread							C_G	C_G					
13. P_S Special possible	R_A		P_G				P_G						

* The table indicates the modality of the conclusion for the (assertorically) valid syllogisms.

** A blank indicates the absence of a valid conclusion.

All data from § 84.

TABLE XII

THE MODALITY OF VALID SYLLOGISMS IN THE THIRD FIGURE

Minor \ Major	1 N_A	2 R_A	3 C_G	4 V_G	5 A_G	6 P_G	7 C_S	8 V_S	9 $\bar{N}_E$	10 $\bar{R}_E$	11 T	12 S	13 P_S
1. N_A Absolute necessary			N_A	N_A			C_S	C_S					
2. R_A Absolute perpetual			R_A	R_A			C_S	C_S					
3. C_G General conditional			C_G	C_G			C_S	C_S					
4. V_G General conventional			V_G	V_G			V_S	V_S					
5. A_G General absolute	Same as entry at head of column		A_G	A_G	Same as entry at head of column		$\bar{R}_E$	$\bar{R}_E$	Same as entry at head of column				
6. P_G General possible			P_G	P_G			P_S	P_S					
7. C_S Special conditional			C_G	C_G			C_S	C_S					
8. V_S Special conventional			V_G	V_G			V_S	V_S					
9. $\bar{N}_E$ Non-necessary existential			V_G	V_G			$\bar{N}_E$	$\bar{N}_E$					
10. $\bar{R}_E$ Non-perpetual existential			V_G	V_G			$\bar{R}_E$	$\bar{R}_E$					
11. T Temporal			C_G	C_G			T	T					
12. S Spread			C_G	C_G			S	S					
13. P_S Special possible			P_G	P_G			P_S	P_S					

NOTE: The table indicates the modality of the conclusion for the (assertorically) valid syllogisms.

All data from § 85.

TABLE XIII

THE MODALITY OF VALID SYLLOGISMS
IN THE FOURTH FIGURE*

Minor \ Major	1 N_A	2 R_A	3 C_G	4 V_G	5 A_G	6 P_G	7 C_S	8 V_S	9 $\bar{N}_E$	10 $\bar{R}_E$	11 T	12 S	13 P_S
1. N_A Absolute necessary													
2. R_A Absolute perpetual													
3. C_G General conditional													
4. V_G General conventional													
5. A_G General absolute													
6. P_G General possible					(SEE BELOW**)								
7. C_S Special conditional													
8. V_S Special conventional													
9. $\bar{N}_E$ Non-necessary existential													
10. $\bar{R}_E$ Non-perpetual existential													
11. T Temporal													
12. S Spread													
13. P_S Special possible													

* The table indicates the modality of the conclusion for the (assertorically) valid syllogisms.

** For the details, which are more complex than with the other figures, see § 86 itself, as given in Appendix A. In general, the procedure is to reduce the fourth figure to one of the three preceding ones.

	Avicenna	*al-Qazwīnī*
	and a judgment is expressed in it without mention of its necessity, nor of its duration, nor of any other temporal circumstance, nor according to possibility	
	or something of this sort is expressed in it: necessity, or duration without necessity, or [actual] existence without either duration of necessity.	§ 48 [the modalized]
(1)	Necessity sometimes occurs in a categorical [or unqualified] sense, as in the statement "God – be he exalted – exists"; and sometimes necessity is (p. 135) attached to a condition. The condition may be *either*	[not considered: the *categorically* or *unqualifiedly* necessary]
(2)	the [entire] duration of the existence of the essence [of the subject]. Thus consider: "Man is necessarily a being that speaks." We do not mean here that man never has ceased nor will cease to be a being that speaks – which would be false of each human being – rather we mean that as long as he remains a human being in his essence, he is a being capable of speech ...	the absolute necessary (§ 50.1)
(3)	*Or* this condition may be the duration of the state in which the subject remains qualifiable by certain qualities which it possesses [i.e., the subject may be such that it must of necessity be subject to the predicate as long as it (the subject) remains under a certain condition] ...	the general conditional (§ 50.3)
(4)	The case is different ... [when] a condition is applicable to subjects [under a limitation of non-perpetuity]	the special conditional (§ 51.1)
(5)	So that it applies (p. 136) neither for a fixed time, as with an eclipse,	the temporal (§ 51.5)
(6)	nor for one that is left indeterminate, as with respiration.	the spread (§ 51.6)

It should be noted that these six modes of necessity include all five of al-Qazwīnī's thirteen modalities which take the form $(A \,/\, \Box \,/\, -)$ with or without the qualification of non-perpetuity $({\sim}A \,/\, \exists t \,/\, E![{\sim}A])$. It warrants note that Averroes' critical discussion states that in the *Kitāb al-shifāʾ* Avicenna treated just these five modes of necessity[17], namely numbers (1)–(6) with a conflation of the first two.

Turning now to the various kinds of possibility, we find the following situation:

	Avicenna	*al-Qazwīnī*
	(P. 138) By "possibility" we understand	
(1a)	either that which goes with the negation of [the fact] that necessity of [its] non-existence – that is to say (p. 139) impossibility in the primary and general sense of the term – is to be (p. 140) attributed to the subject ...	the general possible (§ 50.6)
(1b)	or one understands that which goes with the conjoint negation of the necessity of the non-existence and the necessity of the existence of the subject ... so that it is possible that it be and possible that it not be, i.e., not impossible that it be and not impossible that it not be.	the special possible (§ 51.7)
(2)	(P. 141) ... this [second] sense of "possible" includes the thing whose existence is not perpetual even if it is necessary some of the time, as an eclipse.	the general absolute (§ 50.5)
(3)	The "possible" is sometimes given a third sense. [Here] the judgment is not [strictly] necessary at all – neither at a given time, as with the eclipse, nor under certain circumstances, as with change for that which moves – but it is [actual at some times and not others], as happens with writing for man.	the non-perpetual existential (§ 51.4) and the non-necessary existential (§ 51.3)
(4)	Sometimes one understands it [the possible] in yet another sense. In this case one looks to a relation [of the subject to the predicate] which does not belong to the thing at issue at any as yet realized (actualized) stage of its existence, neither by affirmation nor negation. Rather, one looks to its [the subjects] state in the future.	[not considered: the future contingent]

[17] See 'Averroes' *Quaesitum* on Assertoric Propositions' in N. Rescher, *Studies in the History of Arabic Logic* (op.cit.), pp. 91–105 (see § 11 on p. 103).

It appears that Avicenna's senses of "possibility" correspond in substance to five further members of al-Qazwīnī al-Kātibī's list of thirteen modalities. These are just the five which possess, in our symbolism, either of the forms

$$(A \mid \Diamond \mid -) \qquad \text{or} \qquad (A \mid \exists t \mid -)$$

with or without an amendment of non-perpetuity $(\sim A \mid \Diamond \mid E![\sim A])$ or $(\sim A \mid \exists t \mid E![\sim A])$.

Furthermore in discussing (on pp. 144–145) the modal determinations of the **A** proposition "All *C* is *B*" Avicenna distinguishes three further possibilities, namely that the relationship may be asserted (i) invariably as long as the subject really exists, but without necessity, (ii) always under certain definite circumstances, (iii) at certain times when the subject really exists, though not perpetually. It appears that these three modes correspond to the absolute perpetual (§ 50.2), the general conventional (§ 50.4), and the special conventional (§ 51.2), respectively. These three different modes of actuality – that is, *assertoric* modalities – have the feature of being represented, in our symbolism, by expressions of the form

$$(A \mid \forall t \mid -)$$

with or without an amendment of non-perpetuity $(\sim A \mid \exists t \mid E![\sim A])$.

Thus, not only does Avicenna treat the thirteen modalities of al-Qazwīnī, but he classifies them into three groups, as follows:

Modes of Necessity	Modes of Actuality	Modes of Possibility
Absolute necessary	Absolute perpetual	General absolute
General conditional	General conventional	General possible
Special conditional	Special conventional	Non-necessary existential
Temporal		Non-perpetual existential
Spread		Special possible

Avicenna's arch-critic Averroes, rejects this complicated machinery as unwarranted and unnecessary.[18] Averroes affords one interesting historical datum. According to him, Avicenna maintained in the *Kitāb al-shifā*ʾ that Alexander of Aphrodisias construed necessity to include only the first two of these modes – i.e., the absolute necessary and the general conditional – and that, accordingly class the other three modalities of the first group (viz.

[18] *Ibid.*, §§ 11–13.

the special conditional, the temporal, and the spread) as types of actuality. Be this as it may, Averroes himself gives a – to be sure significantly simplified – temporal construction of modalities, presented in a square of opposition as follows[19]:

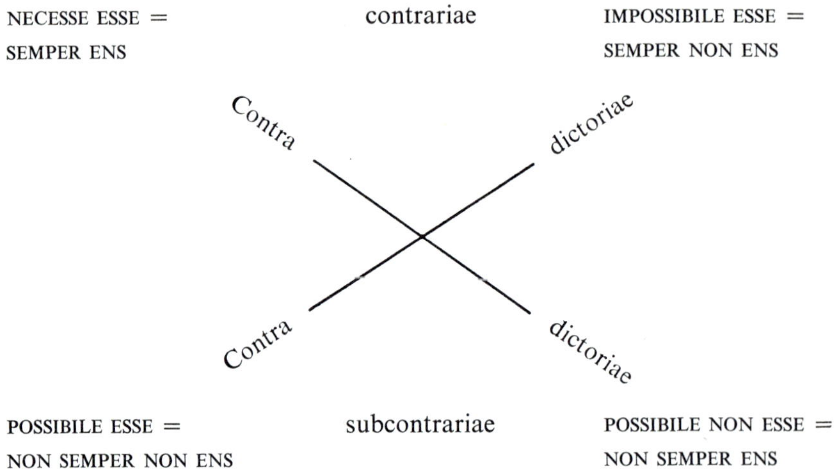

13 TEMPORAL MODALITIES AMONG THE ANCIENT GREEKS AND THE LATIN MEDIEVALS

The early history of temporal quantifiers like "sometimes" and "always", and of the theory of temporalized modalities is that linked to them through the mediation of such principles as "What is sometimes actual is always possible" remains shrouded in obscurity. We know that the rudiments of such a theory was actively developed by the ancient Greeks: the Megarians and the Stoics[20], and Aristotle and the early Peripatetics.[21] The notions of temporalized modality that are at work here are mainly those relating to the "Master Argument" of Diodorus Cronus.[22]

There seems to have been a disagreement as to modality between the Stoics and the Megarians. On the Megarian view:

(1) The *true* is that which is actually realized *now*

$$T_n(p) \text{ iff } R_n(p)$$

[19] Averroes, *In I De Caelo*, t. 5, f. 85A.

[20] See E. Zeller, *Die Philosophie der Griechen*, Pt. 3, Vol. 1 (5th ed., Leipzig, 1923); and B. Mates, *Stoic Logic* (Berkeley and Los Angeles, 1953), see esp. pp. 36–41.

[21] I. M. Bocheński, *La logique de Théophraste* (Freiburg, 1947).

[22] See Mates (op. cit.), pp. 38–39. Cf. J. Hintikka, 'Aristotle and the "Master Argument" of Diodorus', *American Philosophical Quarterly* **1** (1964) 101–114. And see also N. Rescher, 'A Version of the "Master Argument" of Diodorus', *Journal of Philosophy* **63** (1966) 438–445.

or more generally

$$T_t(p) \text{ iff } R_t(p)$$

(2) The *possible* is that which is actually realized (i.e., true) at some present-or-future time

$$P_n(p) \text{ iff } (\exists t)\,[t \geq n \;\&\; R_n(p)]$$

or more generally

$$P_t(p) \text{ iff } (\exists t)\,[t' \geq t \;\&\; R_{t'}(p)]$$

(3) The *necessary* is that which is actually realized at every future time

$$N_n(p) \text{ iff } (\forall t)\,[t \geq n \rightarrow R_n(p)]$$

or more generally

$$N_t(p) \text{ iff } (\forall t')\,[t' \geq t \rightarrow R_{t'}(p)]$$

The Stoics, on the other hand, dropped the now-relativization of the modalities of possibility and necessity, retaining it only for truth (actuality):

(1) The *true* is that which is actually realized *now*

$$T_n(p) \text{ iff } R_n(p)$$

(2) The *possible* is that which is actually realized at *some* (i.e., any) time

$$P(p) \text{ iff } (\exists t)\,R_t(p)$$

(3) The *necessary* is that which is actually realized at *all* times

$$N(p) \text{ iff } (\forall t)\,R_t(p)$$

Aristotle's position is in line with that of the Stoics in viewing the necessary as that which is true all of the time[23], a position faithfully reflected in St. Thomas Aquinas' statement that: *Et sic quidquid semper est, non contingenter semper est, sed ex necessitate.*[24]

In all of this there is no limit of the ramified machinery of temporalized modalities which we find in Arabic texts – and which are unquestionably of Greek provenience. For the roots of this theory we must undoubtedly look to the Stoic doctrine of predication. The Stoics distinguished between three types of qualities:

[23] J. Hintikka (op.cit.).

[24] *In I De Caelo*, lect. 26, n. 258. And correspondingly: *quod possible est non esse, quandoque non est* (*Summa Theologica*, IA, q. 2, a.3). Cf. Guy Jalbert, *Nécessité et contingence chez saint Thomas d'Aquin et chez ses prédécesseurs* (Ottawa, 1961), pp. 204–206, 224–225, and 228.

		i. *poiotēs* (permanent property)
Poion (quality)	{	ii. *schēsis* (enduring state)
		iii. *hexis* (transient characteristic)

In construing "quality" (*to poion*) here, we are to work from the top down, and thus have three possibilities[25]:

(1) Only group (i): those qualities that are wholly completed and altogether permanent (*apartizontas kai emmonous ontas*).

(2) Groups (i) and (ii): not only the permanent qualities (e.g., a man's "being an animal") but the enduring states as well (e.g., "being prudent").

(3) Groups (i)–(iii): adding to (2) also strictly transient qualities (e.g., "walking" or "running").

The distinction between such types of qualities lends itself readily to temporalization in the interpretation of propositions in which they are attributed

A man is an animal *all of the time.*
A prudent man acts wisely *most of the time.*
A healthy man walks *some of the time.*

The elaboration of such temporalized predications would lend itself naturally to the development of temporal modalities of the sort we have here investigated on the basis of discussions in Arabic logical texts.

The Latin medievals seem to have taken this up from the Arabs to a very modest extent. Thus we find in Pseudo-Scotus a distinction between four types of temporalized "necessity" (*conditionale* = conditional, *quando* = as-long-as, *ut nunc* = as-of-now, and *pro semper* = for all times).[26] I would conjecture a correspondence with the cognate Arabic ideas along something like the following lines:

Pseudo-Scotus	Construction	al-Qazwīnī
1. *quando*	$(A \mid \Box \mid E![A])$	absolute necessary
2. *pro semper*	$(A \mid \forall t \mid E![A])$	absolute perpetual
3. *conditionale*	$(A \mid \Box \mid C[A])$	general conditional
4. *ut nunc*[27]	$(A \mid \exists t \mid E![A])$	general absolute

If this conjecture – or anything like it – is correct, temporalized modalities also found made their way pretty well intact into the Latin scholastics. While this development could have been mediated (or perhaps only strengthened) through Arabic materials, it could also have been indigenous to a purely

[25] I follow E. Zeller (op.cit.), pp. 97–99 (especially no. 1 for p. 97); relying also upon Émile Bréhier, *La théorie des incorporels dans l'ancien Stoïcisme* (2d ed., Paris, 1928), p. 9.
[26] I. M. Bocheński, o.p., *Notes historiques sur les propositions modales* (Quebec, 1951), p. 7.
[27] We here construe "*nunc*" not as *the now* (that is, the now-of-the-present), but as *a now* (that is, some – i.e., *any* – instant). This dual interpretation is standard in the medievals.

Latin tradition. Signs of this are already to be found in Boethius: "ea vero quae ex necessitate aliquid inesse designat tribus dicitur modis; uno quidem quo ei similis est propositioni quae inesse significat ... alia vero necessitatis significatio est, cum hoc modo proponimus "hominem necesse est cor habere, dum est atque vivit" ... alia vero necessitatis significatio est universalis et propria quo absolute praedicat necessitatem ... possible autem idem tribus dicitur modis; aut enim quod est, possibile esse dicitur ... aut quod omni tempore contingere potest, dum ea res permanet cui aliquid contingere posse proponitur ... item possibile est quod absolute omni tempore contingere potest. ... ex his igitur apparet alias propositiones esse inesse significantes alias necessarias alias contingentes atque possibiles, quarum necessariarum contingentiumque cum sit trina partitio, singulae ex iisdem partionibus ad eas quae inesse significant referentur; restant igitur duae necessariae et duae contingentes quae cum ea quae inesse significat enumeratae quinque omnes propositionum faciunt differentias; omnium vero harum propositionum aliae sunt affirmativae aliae negativae."[28] Cognate distinctions are also drawn elsewhere. For example, in St. Thomas Aquinas and other medieval scholastics we find a distinction – akin to the modern distinction between logical and physical necessity – between those relationships which are perpetual and (in a sense) necessary: (1) by an eternity that is *a parte ante*, or (2) by an eternity that is *a parte post*. Truths of the former class are necessary by a necessity that turns wholly on the nature of the essences involved – as men are eternally rational and equiangular triangles eternally equilateral. Truths of the second class are necessary by a necessity that devolves from the *au fond* contingent arrangements of this world – as men are eternally mortal, or the northern latitudes eternally cold.[29] Traces of such an interest in temporal modalities are to be found as late as William of Ockham.[30]

14 CONCLUSION

We have seen that ideas of temporalized modality originating in ancient Greece were carried forward and elaborated with a high degree of sophisti-

[28] Quoted from C. Prantl, *Geschichte der Logik im Abendlande*, vol. I (Leipzig, 1855; photoreprinted, Graz, 1955), p. 703, no. 150.

[29] See Guy Jalbert, op. cit., pp. 41, 119–120, 137–138, 141–143. This work is primarily concerned with possible and necessary *existents*. For a detailed treatment of such existents, primarily in Avicenna, but with some comparisons and contrasts in St. Thomas, can be found in Gerard Smith, 'Avicenna and the Possibles', *The New Scholasticism* **17** (1943) 340–357.

[30] *Summa logicae* (ed. P. Boehner), Pt. I, ch. 73, lines 16–49; Pt. II, ch's 7, 19–22; Pt. III, div. i, ch's 17–19.

cation by Arabic logicians of the middle ages. The intricacy of the modal notions at issue was such that, in the context of complex interrelationships – and especially in the modal syllogistic – errors tended to creep in. Nevertheless, the basic ideas were clear and the guiding intuitions of their elaboration straightforwardly intelligible. It is, I believe, safe to say that – barring possible new findings from the side of medieval Latin scholasticism – the logical theory of temporal concepts was carried to a higher point in Arabic logic than at any subsequent juncture prior to our own generation.

APPENDIX A

THE HERETOFORE UNTRANSLATED SECTIONS OF AL-QAZWĪNĪ AL-KĀTIBĪ'S *AL-RISĀLAH AL-SHAMSIYYAH*

INTRODUCTION

As already mentioned in Section 2 of the monograph, the Arabic text of "The Sun Epistle" (*Al-Risālah al-shamsiyyah*) of al-Qazwīnī al-Kātibī was published by Aloys Sprenger over a century ago: Aloys Sprenger, *Dictionary of the Technical Terms Used on the Sciences of the Musulmans*, Part 2 (Calcutta, 1862.) See Appendix I on 'The Logic of the Arabians'. This publication also presents an English translation of the work, which, however, omits the three key sections devoted to the logic of modality. The aim of the present Appendix is to close this gap by presenting a translation of this omitted material. Our translation is based on the Arabic text published by Sprenger, a few minor departures from which (indicated in the footnotes) have proved to be necessary.

The paragraphs have been numbered dually. The *second* number is that of Sprenger's Arabic text and the *first* number fits the paragraph into the numbering of the printed English translation (whose numbering is unaccountably out of step with that of the Arabic text).

Second Book/Third Section/Second Inquiry:

ON SIMPLE [OR: EVEN] CONVERSION (CONVERSIO SIMPLEX)

(Continuation)

[Negative Particulars]

(68/66) If it [the original proposition] is particular [negative], then the special conditional and the special conventional are both convertible to the special conventional. [This is so] because if it is true by necessity or perpetually that *Some C is not B so long as C exists [but] not perpetually,* [then] it is true perpetually that *Some B is not C so long as B exists [but] not perpetually.* For let us suppose the essence (or: *nature*) of the subject, namely *C*, to be *D*. Then *D* is *C* in actuality (or: *in fact*), and *D* is *B* also, because of the non-perpetuity of the denial of *B*-ness of it (viz. *D*). It (viz. *D*?) is not *C* so long as *B* exists; otherwise it (viz. *D*) [which is both *B* and *C*, as just seen] will be *C* while it is *B*, and therefore *B* while it is *C*. But it is already [supposed] that it (viz. *C*) is not *B* so long as *C* exists. This is absurd. And if it is true that *C*-ness and *B*-ness [both] belong to it (viz. *D*) [at some times] and denied of it [at others], then it is true that *Some B is not C so long as B exists but not perpetually*. And this is the sought-for [conclusion].

As to the rest [of the negative particulars], they are not convertible, because it is true by necessity that *Some animals are not men*, and by necessity that *Some moons are not eclipsed at the time of the quarter-moon but not perpetually* [then] their converses are false by a general impossibility. However, the necessary [proposition] specializes the elementaries, and the temporal specializes the remaining linkages; and when they [viz., the specials] are not convertible, none of them [viz., the general] are convertible. [For] you know that the convertibility of the general necessitates the nonconvertibility of the special.

[Affirmatives]

(69/67) As to the affirmative, be it either universal or particular [the situation is the same, since] a universal original [proposition] does not convert[1], because of the possibility of the existence of a predicate more general than the subject, as in the statement *Every man is an animal.*

As to [the affirmative universal or particular] with modality, the necessary and the perpetual and the two generals [viz. the general conditional and the general conventional] are convertible to the absolute temporal (sic!) because if it is true that *All C is B* by one of these four aforementioned modalities, then *Some B is C while it is B*. Otherwise, *No B is C so long as B exists*; and this, together with the original [viz., *All C is B*], yields *No C is C necessarily or perpetually* in [the case of] the necessary and the perpetual, and while it (i.e., *C*) is *C* in [the case of] the two generals. And this is absurd.

As to the two specials [viz. the special conditional and the special conventional], they are convertible to the absolute temporal (sic!) bounded by the non-perpetual [i.e., by the restriction of non-perpetuity].[2] As to the absolute temporal, its existence is a requisite for their generality. And as to the bound of the non-perpetual: (i) in [case of] a universal [affirmative] original, it obtains because if it were falsified that *Some B is not C in actuality* (or: *in fact*) by the truth that *All B is C perpetually*, then we connect it with the first part of the original, namely the statement [true] by necessity or perpetually that *All C is B so long as C exists*, yielding *All B is B perpetually*. And then we connect it with the second part also, namely the statement that *No C is B by*

[1] That is, it does not convert strictly, but only by limitation.
[2] This, then, would be a special conditional.

a general absolute, yielding *No B is B by a general absolute*. This necessitates a pair of contradictories, and this is absurd. (ii) As to [the case of] the [affirmative] particular [original] [the bound of the non-perpetual also obtains]. Thus we shall suppose the subject to be D[3], and that it is not C in actuality (or: *in fact*), for otherwise it will be C perpetually, and so B perpetually, because of the perpetuity of B-ness with the perpetuity of C-ness. But then it is useless for limiting an original [proposition] by the non-perpetual.[4]

As to the two temporals [viz. the temporal and the spread] and the two existentials [i.e., the non-necessary existential and the non-perpetual existential] and the general absolute, they convert to the general absolute. [This is so] because if it is true that *All C is B* by one of these aforementioned five modalities, then *Some B is C by a general absolute.* Otherwise it is true that *No B is C perpetually*, and this, together with the original yields *No C is C perpetually*, and this is absurd. If you want to convert the opposite (or: *contrary*) of the converse in the two affirmatives [viz., universal or particular], then make-true the opposite (or: *contrary*) of the original, or a particularization (or: *specialization*) of it.

(70/68) As to the two possibles, their state in [regard to] [both] conversion or its nonexistence is unknown. [This is] because of the stoppage of the aforementioned demonstration (or: *proof*) of convertibility, since convertibility in these two [cases] [obtains]: (i) according to the conversion of the negative necessary as such, or (ii) according to the yielding [of a conclusion] by a possible minor together with a negative major in the first figure and the third, both of which are uncertain. By the nonexistence of success in the demonstration there is necessitated [the possibility of] [both] conversion and its nonexistence.

[Conditional Propositions]

As to the conditional, the affirmative "conjunctive" [proposition] converts to a particular affirmative, and the negative universal to a negative universal. For if the contradictory of a converse were true, it would not connect with the original [to yield] a syllogism that would yield absurdity. As to the negative ["conjunctive"] particular, it is not convertible. [This is] because the truth of the statement *It is not so that if this is an animal then it is a man* [obtains] together with the falsity of its converse.

And as to the [second type of conditional proposition, viz. the] "disjunctive", they do not admit of forming a converse because of the nonexistence of a natural differentiation between their parts.

[3] So the text, which seems corrupt at this point.
[4] That is, the absolute perpetual is not subject to the limitation of non-perpetuity.

Second Book/Third Section/Third Inquiry:

ON CONVERSION BY CONTRADICTION

[Affirmative Universal Case]

(72/70) As to the affirmatives, if they are universal, then seven of them – namely those whose negatives are not convertible by a simple (or: *even*) conversion (*conversio simplex*)[5] – are not convertible; because it is true by necessity that *Every moon is not eclipsed at the time of the quarter-moon but not perpetually* [*i.e., Some moon is eclipsed at the time of the quarter-moon, but not perpetually*] without its [contradictory] converse being something you know [to be true].

The necessary and the perpetual [universal affirmative] are convertible to a perpetual universal [negative], because if it is true by necessity or perpetually that *All C is B*, then perpetually *Nothing that is not B is C*. Otherwise, *Something that is not B is C in actuality* (or: *in fact*); and this, together with the original [proposition] yields *Something that is not B is B by necessity* in [the case of] the necessary and *perpetually* in [the case of] the perpetual. And this is absurd.

As to the universal [affirmative] [general] conditional and [general] conventional, both are convertible to a general conventional universal. [This is so] because if it is true by necessity or perpetually that *All C is B as long as C exists*, then perpetually *Nothing that is not B is C so long as it is not B*. Otherwise, *Something that is not B is C while it is not B*. This, together with the original, yields *Something that is not B is B while it is B*, and this is absurd.

As to the two specials [viz., the special conditional and the special conventional], they are both convertible to the general conventional non-perpetual [and thus the special conventional], about some [i.e., particular].

As to [its being] general conventional, this is because of its being necessitated by either of the two generals. As to [its being] non-perpetual about some [i.e., particular], this is because it renders true that *Something that is not B is C with general absoluteness*. Otherwise, *Nothing that is not B is C perpetually*, which converts to *Nothing that is C is not B perpetually*. But it is already [supposed] that *Nothing that is C is B in actuality* (or: *in fact*) by a judgment of non-perpetuity in the original [statement]. This necessitates *All C is not B in actuality* (or: *in fact*), by the existence of the subject. This is absurd.

[5] These seven are enumerated in § 65.

[Affirmative Particular Case]

(73/71) And if it [the affirmative modal proposition] is particular, then the two specials are convertible to the special conventional. [This is so] because if it is true by necessity or perpetually that *Some C is B so long as C exists [but] not perpetually*, then it is requisite that it be true that *Something that is not B is not C so long as it is not B [but] not perpetually*. [This is so] because we suppose that the essence [nature] of the subject – namely *C* – is [something, say] *D*; and so *D is not B in actuality* (or: *in fact*) [though] not perpetually, due to the non-perpetuity of the constancy (or: *unchangeability*) of its *B*-hood, and it (viz. *D*) is not *C* so long as it is not *B*. Otherwise, it (i.e., *D*) will be *C* while it is not *B*, and so not *B* while it is *C*. But it is already [held to be] *B* as long as *C* exists. This is impossible. And *D* is *C* in actuality (or: *in fact*). This is self-evident [since *D* is the essence of *C*]. Thus *Something* (viz. *D*) *that is not B is not C so long as not B exists but not perpetually*. And this is the sought-for [conclusion].

As to the rest [of the affirmative particular modals], they are not convertible, since the truth of the statement *Some animals are not men, by an absolute necessary* and [the statement] *Some moons are not eclipsed, by a necessary temporal*[6] [obtains] without the converse [being true] by [any of] the whole group of the modalities. And when they [i.e., the necessary absolute and the temporal] are not convertible, no single one of them is convertible, as you know respecting simple (or: *even*) conversion (*conversio simplex*).[7]

[The Negative Cases: Universal and Particular]

(74/72) As to the negatives, they are either universal or particular. The universal is not convertible (i.e., *C*-convertible), due to the possibility that the extension [literally: *the being*] of the opposite (contrary) of the predicate is more general than the subject.

[As to the particulars], the two specials convert to the temporal absolute (sic!). [This is so] because if it is true by necessity or perpetually that *Nothing that is C is B so long as C exists [but] not perpetually*, we suppose that the subject is *D*, so that it (i.e., *D*) is not *B* in actuality (or: *in fact*), but *D* is *C* at some times that it is not *B*, because it is not *B* at all the times that it is *C*. Thus *Something* (viz. *D*) *that is not B is C at some time when it is not B*. And this is what was called for [as to be shown].

As to the two temporals and the two existentials (or: *actuals*), they are convertible to the general absolute. [This is so] because if it is true that *Nothing that is C is B* by any one of these modalities, we [again] suppose the

[6] The text elsewhere knows nothing of a "necessary temporal".
[7] See § 68, *ad fin.*

subject to be *D*, so that it (i.e., *D*) is not *B* in actuality (or: *in fact*), and it is *C* in actuality (or: *in fact*), because of the [current] existence of the subject. Thus, something (viz. *D*) that is not *B* is *C* in actuality (or: *in fact*), and this is the sought-for [conclusion]. In this way the conversions of their particulars [sc. of the temporals and existentials] is established.

As to the rest of the negatives – and the conditional, be it either affirmative or negative – their convertibility is unknown, due to the lack of a successful demonstration [literally: *due to the lack of success with the demonstration*].

Third Book/Second Section:

ON MIXED [SYLLOGISMS]

(83[8]/81) As to the first figure, its condition (for validity) [obtains] in relation to the modality operative for the minor. The conclusion here is the same as the major [in modality], if it (i.e., the minor) is other than one of the two conditionals[9] and the two conventionals[10]; and otherwise [i.e., if the minor *is* one of these four], then [the modality of the conclusion is] the same as the minor, removing from it (i.e., the conclusion) the restriction (or: *limitation*) of the non-necessary or the non-perpetual[11] of the minor, if the major is one of the two generals, and after adding the non-perpetual to it (i.e., the conclusion) if it (i.e., the major) is one of the two specials.

(84/82) As to the second figure, its condition (for validity) in relation to modality [is determined] by two matters: (i) One of the two is that truth by perpetuity [obtains] for the minor, or the occurrence (or: *existence*) of the major in one of the two convertible negative propositions.[12] (ii) The second is that the possible is not used [in the premisses] except with the absolute necessary, or with one of the two conditional major premisses. The conclusion is perpetual if truth by perpetuity [obtains] for one of the two premisses; and if not, then [the modality of the conclusion] is as the minor, removing from it the restriction (or: *limitation*) of the non-perpetual and the non-necessary, and of the necessary, whatever [mode of] necessity it is.[13]

[8] The Sprenger translation unaccountably labels the last sentence of the preceding section as a separate § 83.
[9] Namely the "general conditional" and the "special conditional". Cf. above, §§ 50–51.
[10] Namely the "general conventional" and the "special conventional". Cf. above, §§ 50–51.
[11] Here the text, which elsewhere knows nothing of a "special necessary", adds "and the special necessary". This (apparent) corruption must be suppressed.
[12] Namely the "special conditional" and the "special conventional".
[13] Are there more than one? Compare the last sentence of the preceding paragraph, especially its last footnote. In any event, the purport of the clause beginning "and of the necessary ..." is unclear (to me).

(85/83) As to the third figure, its condition (for validity) [obtains] [in relation to the modality operative] for the minor. The conclusion is as the major, if the major is other than the four [kinds][14]; and otherwise it is as the minor[15] removing the non-perpetual from it if the major is one of the two generals, and adding [the non-perpetual] to it (the minor) if it (the major) is one of the two specials.

(86/84) As to the fourth figure, the condition for its validity in relation to modality [is determined] by five matters: (i) The first is the occurrence [or: *existence*] of the syllogism in it [composed] of two actuals.[16] (ii) The second is the conversion of the usable negative in it.[17] (iii) The third is that truth by perpetuity pertains to the minor in the third mood, or that the general conventional pertains to the major. (iv) The fourth is the occurrence [or: *existence*] of the major in the sixth [mood] in the negative conversions.[18] (v) The fifth [and last] is the occurrence [or: *existence*] of the minor in the eighth [mood] in one of the two specials, the major being true by a general conventional.

The conclusion is as follows: (1,2) In the first two moods [a conclusion is obtained] by conversion of the minor if it is a perpetual truth, or if the minor[19] is [subject to] the six negative conversions, and otherwise it (i.e., the conclusion) is a general absolute. (3) In the third mood it (i.e., the conclusion) is perpetual if one of the two premisses is true perpetually, and if not, then the minor is converted. (4,5) In the fourth and fifth [moods], it (i.e., the conclusion) is perpetual if the major is true perpetually, and if not, then the minor is converted, removing the non-perpetual from it. (6) In the sixth [mood], it is the same as in the second figure after conversion of the minor. (7) In the seventh [mood] it is the same as in the third figure after conversion of the major. (8) In the eighth [mood] [the case is] as though one converts the conclusion after changing [literally: *converting*] the order [of the two premisses].

[14] Viz. two generals and two specials, as the rest of the paragraph shows: specifically the general conditional, the general conventional, the special conditional, and the special conventional.

[15] The text reads "as though the minor is converted" instead of "as the minor". I propose to amend it to read exactly as at the end of the preceding paragraph.

[16] That is, here, as elsewhere, only assertorically valid syllogisms come into question for modal validity.

[17] What does this *mean*?

[18] The "negative conversions" are presumably the six modalities in which E propositions are convertible, as listed in §§ 65–67, viz. the absolute necessary and perpetual, and the general and special conditionals and conventionals.

[19] The text reads: "the syllogism".

APPENDIX B

A FRAGMENT OF GALEN'S LOST TREATISE "ON POSSIBILITY"

I. INTRODUCTION

It has been known for over two centuries that there exists in the fine collection of Arabic manuscripts in the Escorial library near Madrid a group of Arabic translations of treatises by Alexander of Aphrodisias, including his refutation of Galen's treatise on possibility. These treatises are reported in Casiri's catalog of 1760[1], and figure in many later bibliographies.[2]

The translation we present below is made from the aforementioned unicum manuscript of the Escorial of Alexander of Aphrodisias' 'Refutation of Galen on the Possible'.[3] However, the composite nature of the Escorial manuscript is clearly indicated by the fact that its *incipit* reads: "The treatise of Alexander of Aphrodisias in refutation of Galen's discourse on the possible" whereas its *explicit* reads "Alexander's treatise in reply to Galen concerning the first mover is ended". In fact, the first page alone deals with possibility (only) and the whole (self-contained) remainder – written in the same hand – deals with the theory of motion. It is clear that the Escorial manuscript, rather than presenting a composite treatise, is unquestionably such that two distinct but juxtaposed works were somehow shuffled together (a thing by no means impossible in view of the casual conditions which prevailed at the Escorial for many years in the handling of Arabic manuscripts[4]). The beginning of the treatise on motion is clearly missing, and also – alas – all but the initial folio of the treatise on possibility.

The Escorial manuscript of Alexander's treatise on Galen on the first mover contains at its outset a single manuscript page (folio 59 *verso*) of

[1] Michael Casiri, *Bibliotheca Arabico-Hispana Escurialensis*, 2 vols. (Madrid, 1760, 1770); see Vol. I, p. 242, codex no. 794 (the Galen refutation is part of the second part of the codex, folios 59–69). In the later catalog of Derenbourg/Renaud (*Les manuscrits arabes de l'Escurial*, Paris, 1941), this codex becomes no. 798.

[2] J. G. Wenrich, *De auctorum graecorum versionibus et commentariis* etc. (Leipzig, 1842), p. 276. L. Leclerc, *Histoire de la médecine arabe*, vol. I (Paris, 1876), pp. 216–217; M. Steinschneider, *Die Arabischen Uebersetzungen* ..., pp. 93–97, and his earlier *Al-Fārābī* (St. Petersbourg, 1869), p. 93.

[3] The scribe of the Escorial MS is Abū 'l-Ḥunain ibn Ibrāhīm ibn Yazīd al-Kātib. We know nothing else of him. His name is given as the scribe of the initial page of our MS.

[4] See the introductory remarks of Derenbourg's catalog. Indeed it is not inconceivable that the remainder of the treatise on possibility will yet turn up in the Escorial, somehow misfiled.

Alexander's treatise on Galen on possibility. This one page – the Arabic text of which has been presented elsewhere[5] – consists almost entirely of a single long quotation from Galen's (lost) treatise *De possibilitate*. This is the only portion of this treatise that is known to have survived.[6]

The Galen-fragment at issue is, in fact, of considerable interest for the history of logic. It is well-known that Aristotle (and some of the Stoic logicians) proposed to construe the modality of *necessity* in terms of *omnitemporality* (construing, for example, "Water is *necessarily* wet" as "Water is wet *at all times*"). And as we have seen, the Arabic logicians carried this temporal construction of modality to far greater lengths, elaborating with great sophistication upon the fundamental Aristotelian linkage of modality to time, based on the idea that the necessary is that which happens all of the time.

Now the surprising fact to be gleaned from Galen's report in the fragment here at issue is that *this development of a ramified linkage between time and modality heretofore encountered no earlier than in Arabic logical texts goes back to the earliest Peripatetics*, i.e., to Theophrastus and his followers, who distinguished exactly between rigidly Aristotelian *omnitemporal* necessity (of the "Water is necessarily wet" type described above) and al-Qazwīnī al-Kātibī's "absolute necessity" (of the type: "A man is necessarily an animal" = "A man is an animal so long as he exists.").[7] This early linkage of time and modality is a fact of which we have heretofore had only somewhat indirect hints, such as that of the following passage from Alexander's *Commentary on 'Prior Analytics'*: "For according to him [Aristotle] the necessary is also predicated of the actual (or: *existent*), for what is actually (or: *existentially*) true of something necessarily belongs to (literally: *exists in*) it, so long as it exists. Thus Theophrastus in the first Book of his *Prior Analytics*, speaking of the meanings of the necessary, writes as follows: 'The third meaning is the existent: for when it exists, it is impossible for it not to exist' (fr. 58 F. Wimmer). (Ed. M. Wallies, 156:27–157:2.)

In this way, our Galen fragment offers a new datum of considerable interest for the history of modal logic.

[5] N. Rescher and M. E. Marmura, *The Refutation by Alexander of Aphrodisias of Galen's Treatise on the Theory of Motion* (Karachi, 1967; Publications of the Central Institute of Islamic Research).

[6] See the survey of Arabic Alexander MSS in Albert Dietrich, *Die Arabische Version einer unbekannten Schrift des Alexander von Aphrodisias über die Differentia specifica, Nachrichten der Akademie der Wissenschaften in Göttingen: Philologisch-historische Klasse*, Jahrgang, 1964, no. 2 (see p. 96, item 11). Here the references to the treatise by the Arabic biobibliographers are also listed.

[7] For a comparable distinction absolutely perpetual motion ("of the eternal everlasting things") and the non-absolutely perpetual motion ("of things which are not eternal and everlasting but which continue to move as long as they return their nature") see the treatise on the prime mover, E68a12–14.

II. TRANSLATION

The Treatise of Alexander of Aphrodisias in Refutation of Galen's Discourse
59bl on the Possible. / Transcribed by Abū 'l-Ḥunain ibn Ibrāhīm ibn Yazīd al-Kātib.

Galen said:

The meaning that is understood by the expression "necessary", though present in both statements together, / that is, in the statement "The sun shines necessarily" and also the statement "Man is rational necessarily" is one and the same. This is so because the expression "necessary" in both these statements refers to no other thing than [the fact] that light exists [invariably] for the sun and rationality for man, an existence which is inseparable. The followers of Theophrastus erred in differentiating between the two things posited as subjects in the two premisses, "transferring" to the unknown[8] the predicate / in both things.[9] Since the sun is eternal, [and so also its light], whereas rationality in [man] becomes corruptible by reason of his corruptibility, they [i.e., the followers of Theophrastus] imagined that the necessary has therefore two meanings.[10] In this, their state resembles the state of someone who sees that "the sensitive" is at one time predicable of the short, at another of the tall, then of the great, the small, the hardy, the weak, and imagines that sensation differs according to the differences between these things so as to say that some sensation is at one time tall, / some at another time short, some hardy, some weak, some great, some small. Thus just as one who divides sensation along these lines has wrought a bad thing by not investigating the expression "sensation" to see in what meaning or meanings it refers to these things – saying that it refers to many diverse things in as much as the meanings it refers to are found in many diverse things – similarly, whoever[11] divides / the necessary into something which is eternal and into something which exists as long as [the thing it refers to] exists, has wrought a bad thing. For he has taken something which is inseparable and divided it according to the differences among the things wherein it exists.

[8] On "transfer to the unknown" see N. Rescher, *Al-Fārābī's Short Commentary on Aristotle's "Prior Analytics"* (Pittsburgh, 1963), pp. 93–98, 108–111.

[9] Presumably this means that since "necessary" is given two senses, we do not know which of the two is operative in a given predication until we look at the subject to see whether it is temporary or eternal; whereas if "necessary" has but one *modus operandi* we can always know how it is used in a given case.

[10] Also apparently the Stoics. See C. Prantl, *Geschichte der Logik im Abendlande*, vol. I (Leipzig, 1855), pp. 567–568. Actually, Galen holds, the mode of temporal necessity uniform in the two cases, since the predicate pertains always to the subject (throughout its existence – be it temporarily on the one hand or omnitemporally on the other).

[11] Presumably "the followers of Theophrastus". And also the Stoics?

INDEX OF NAMES

SUBJECT INDEX